39 Advances in Biochemical Engineering/ Biotechnology

Managing Editor: A. Fiechter

W0106574

Vertebrate Cell Culture II and Enzyme Technology

With Contributions by
A. F. Bückmann, G. Carrea, D. Dubois,
A. Fiechter, F. K. Gmünder, P. Liras,
J. F. Martín, F. D. Menozzi, A. O. A. Miller,
A. Prokop and M. Z. Rosenberg

With 64 Figures and 19 Tables

Springer-Verlag Berlin Heidelberg GmbH

ISBN 978-3-662-15099-3   ISBN 978-3-540-46133-3 (eBook)
DOI 10.1007/978-3-540-46133-3

© by Springer-Verlag Berlin Heidelberg 1989
Originally published by Springer-Verlag Berlin Heidelberg New York in 1989.
Softcover reprint of the hardcover 1st edition 1989
Library of Congress Catalog Coard Number 72-152360

2152/3020-543210

Federico Parisi † 1989

The Advances in Biochemical Engineering/Biotechnology has received with deep sadness the message of the sudden death of Prof. Federico Parisi who leaves behind many friends and colleagues in many countries. A vast scientific community of biotechnology throughout Europe and other continents is mourning over the loss of our Italia pioneer in biotechnology.

Federico Parisi left us in the middle of an active life. He was the promoter of Italy as a candidate for hosting the forthcoming Congress of Biotechnology in 1993 and acting President of its organizing committee.

We met Federico Parisi first twenty years ago at an IUPAC Commission Meeting representing Italy. His broad expertise was recognized by all members of that body when he proposed the elaboration of the alcohol tables. Under his guidance they were later published by the International Union of Pure and Applied

Chemistry and are still in use worldwide. The work has been a great example of international collaboration and efficient committee work by experts. Federico Parisi supported the commission work constantly and represented his very beloved country at the IUPAC-Symposium many times.

Federico Parisi was a particular good friend of Switzerland where he received part of his scientific education under Prof. de Diesbach at Freiburg University. Born in Naples, he graduated in Industrial Chemistry at the University of Milan and completed his professional education at the Laboratory of Research of the Institute de Angeli (Milan) and the Laboratory Sauter (Geneva) before entering the Research Labs of Distillation, a group of ERIDANIA in 1948.

As a leader of this group from 1957 he developed an active scientific life in applied research and academic teaching at the University of Genoa. He became the prominent member of the Italia biotechnology community. As a Professor of the Genoa University he developed very fruitful activities including publication of several most valuable books and editing the official magazine of the Italien Chemical Society "La Chimica e l'Industria" (from 1970 to 1985). A most relevant review article on "Advances in Lignocellulosics Hydrolysis and the Utilization of the Hydrolyzates" has been published in the last volume of this series (Vol. 38, 1989) informing us of his great interest in bio-energetics.

During the preparatory period for founding the European Federation of Biotechnology (EFB) he promoted the founders efficiently and he was elected as the first representative of his country in the Executive Committee of EFB.

Besides his scientific and professional qualities Federico Parisi impressed all of us by his humanity and great enthusiasm. We remember him as a perfect gentleman and a sincere friend.

Table of Contents

Table of Contents of Volume 34

Metabolic Control of Glucose Degradation in Yeast and Tumor Cells

Armin Fiechter and Felix K. Gmünder
Institute of Biotechnology, ETH-Hönggerberg, CH-8093 Zurich, Switzerland

Regulation of glucose degradation in both yeasts and tumor cells is very similar in many respects. In both cases it leads to excretion of intermediary metabolites (e.g. ethanol, lactate) in those cell types where uptake of glucose is unrestricted (*Saccharomyces cerevisiae*, Bowes melanoma cells).

The similarities between glucose metabolism observed in yeast and tumor cells is explained by the fact that cell transformation of animal cells leads to inadequate expression of (proto-)oncogenes, which force the cell to enter the cell cycle. These events are accompanied by alterations at the signal transduction level, a marked increase of glucose transporter synthesis, enhancement of glycolytic key enzyme activities, and slightly reduced respiration of the tumor cell. In relation to homologous glucose degradation found in yeast and tumor cells there exist strong similarities on the level of of cell division cycle genes, signal transduction and regulation of glycolytic key enzymes.

It has been demonstrated that ethanol and lactate excretion in yeast and tumor cells, respectively, result from an overflow reaction at the point of pyruvate that is due to a carbon flux exceeding the capacity of oxidative breakdown. Therefore, the respiratory capacity of a cell determines the amount of glycolytic breakdown products if ample glucose is available. This restricted flux is also referred to as the respiratory bottleneck. The expression "catabolite repression", which is often used in textbooks to explain ethanol and acid excretion, should be abandoned, unless specific mechanisms can be demonstrated.

Furthermore, it was shown that maximum respiration and growth rates are only obtained under optimum culture conditions, where the carbon source is limiting.

Advances in Biochemical Engineering/
Biotechnology, Vol. 39
Managing Editor: A. Fiechter
© Springer-Verlag Berlin Heidelberg 1989

1 Introduction

Sugars have a profound influence on the metabolism of man and microbes. They provide the main source of carbon and energy for cell physiology and proliferation. Among the mono- and polysaccharides, glucose plays a central role in the living world due to its widespread utilization. It is the substrate par excellence and represents the starting point of the major metabolic pathways.

Control mechanisms of glucose degradation have therefore been the focus of research since Pasteur [86] observed reduced ethanol formation in yeast in the presence of oxygen ("Pasteur-effect"). Furthermore, dominance of glucose uptake over other sugars and formation of certain metabolites became known as the "glucose effect". This phenomenon is also known as "catabolite repression", which refers to first, the inhibition of degradation of other substrates by the breakdown products of glucose, and second, reduced respiration when glucose serves as the carbon source. However, Fiechter et al. [32] have shown that the control of metabolic pathways underlying this "glucose effect" occurring in glucose sensitive yeasts has never been elucidated. It is clear that "fermentation" (or excretion of ethanol) by yeasts or formation of lactate by animal cells was related to glucose and oxygen consumption rates or fluxes. Thus, it was realized that the control of respiration became a key aspect in the control of glycolysis.

In the past, regulation of glucose metabolism was studied on the level of substrate or metabolite concentrations instead in terms of turnover rates e.g. metabolite kinetics. Therefore, methodological concepts remained inadequate for decades and even the introduction of "defined" systems using respirometer measurements with resting cells turned out to be artifactual with respect to the in vivo situation of growing cells.

In more recent years, progress has been made in studying the regulation of glucose metabolism due to the improvement of cultivation methods for microbial and animal cells. Growing cell populations were investigated instead of resting cells using chemostat methods. The use of chemostat experimentation including pulse and shift technique has led to new insights into the coordination of metabolic activity as in these cases the growth rate and culture conditions are well defined. Thus, reconsideration of old problems like "catabolite repression" during glucose degradation or regulation of respiration during growth have become more promising. Progress made in animal cell culture in particular will allow a better comparison of glucose metabolism to be made between higher and lower eukaryotes.

Recently, the regulation of glucose metabolism in *Saccharomyces*-type yeasts has been studied using feed-controlled systems [54, 55, 112] as well as ^{13}C NMR and ^{14}C radioactive labelling techniques in accordance with the Warburg manometric methods. In particular, its level of aerobic and anaerobic glycolysis were determined, and the kinetics of yeast phosphofructo 1-kinase (PFK-1) was studied in vitro [25, 26, 97]. More recently, our knowledge of regulation of glycolysis in yeast has been widened by studying the involvement of fructose 2,6-biphosphate in this process [19, 121]. Today, all data available on this metabolic pathway firmly suggest that its regulation is closely linked with the cell's respiration capacity. Thus, under condition of excess glucose uptake rates, the respiratory capacity of the cell is exceeded, and ethanol formation begins [54, 55, 112]. This observation led to the formulation of a model, which explains this overflow of glycolysis products at the point of pyruvate, by the respiratory bottle-neck [55].

In contrast to yeast cells, much less is known regarding the control of glycolysis in tumor cells. Most of these cells consume considerably more glucose than their normal ancestors do. This increased glycolytic activity of tumor cells is a well established phenomenon since the time of Warburg [127]. Moreover, they secrete large amounts of lactate, even in oxygen saturated media, where respiration should be optimal.

This increase of aerobic glycolytic activity in tumor cells may presently be explained by, first, an increased number of glucose transporters [10, 34], second, an increased level of mitochondrial hexokinase (HK) isoenzyme [29, 135], and third, a decrease in the number of mitochondria in tumor cells, leading to ineffective oxidation [29, 87].

A comparison of the metabolic control of glycolysis in yeast and tumor cells allow us to present an updated view of this process.

2 Historical Perspectives

Pasteur studied "fermentation" and the formation of ethanol by yeast. A full understanding of regulation of glycolysis, respiration and growth, however, used to be a subject of confusion and controversy as Pasteur did not measure glucose uptake or growth rates. Instead, he estimated ethanol formation using ill defined systems containing growing cells and excessive sugar. Pasteur reported in 1861 [86] that the efficiency of "fermentation" decreases in the presence of air. This observation was termed the "Pasteur effect" by Warburg [129] in relation to his studies on glucose metabolism in tumors using his newly developed manometric method. Warburg determined lactate and carbon dioxide production in carbohydrate-utilizing cells and tissues. Concentrations of the end products of glycolysis suggested that tumor cells possess a high glycolytic activity and low respiration in comparison with normal cells [127]. Warburg concluded that low respiration is the origin of cancer [127, 129]. This startling yet false conclusion was maintained by the Warburg-school for more than 30 years and led to endless controversies between supporters and opponents [15, 92, 93, 130, 131, 133]. High glycolysis and low respiration is also observed in a variety of normal proliferating and resting cells that have either few or no mitochondria [29].

In contrast to Warburg, Crabtree noticed a slight reduction (10 %) of the respiration in various tumor cells [22]. Crabtree concluded that in spite of a relatively high respiration, this is ineffective in reducing the aerobic glycolysis. He also noted that the glycolytic activity of tumors exerts a regulatory effect on their respiration. However, Crabtree observed that respiration and carbohydrate metabolism varied widely between tumor tissue of different origin. Nevertheless, increased glycolysis was always observed in tumor cells [87, 133]. Thus, Crabtree stated that respiration is unable to affect the glycolytic rate. This is in contrast to Pasteur's observations, which show that respiration does influence the rate of glycolysis.

3 Regulation of Glycolysis

3.1 Yeast

Glucose appears to be the predominant sugar involved in regulating glycolysis and respiration. In the presence of glucose many strains of microbes typically show reduced cell synthesis (biomass yield), respiration, activity of enzymes and excretion of metabolites derived from pyruvate. If glucose and other carbohydrate sources are used simultaneously as a carbon and energy source, glucose is always preferentially degraded. Thus, any model has to explain the interrelationship between glucose uptake, glycolysis and respiration. Taken together, one has to deal with a rather intricate situation which cannot be explained just by the current textbook explanation of "catabolite repression".

3.1.1 Early Experimental Data

The first major hypothesis for regulation of glycolysis in yeast was proposed by Johnson [51] and Lynen [71] in 1941. Johnson regarded reduced glycolysis, as observed by Pasteur, as being due to decreased concentrations of inorganic phosphate (P_i), AMP and ADP. Lynen [71] proposed that glycolysis and oxidative phosphorylation compete for P_i. In the presence of air respiration succeeds and reduces the availability of P_i for generation of ATP at the substrate level. This idea of a key role of P_i in regulating carbohydrate metabolism was further supported by investigations where the Pasteur effect was eliminated by uncoupling the respiratory chain and oxidative phosphorylalation [72]. It was believed that this leads to increased levels of P_i, AMP and ADP that in turn favors high rate of glycolysis. This view is further supported by the observation that glucose uptake and ethanol excretion were augmented.

In the following 10 years, Lynen discovered that under aerobic conditions, where glycolysis is reduced but respiration is increased, the activity of hexokinase (HK) is low. Since the activity of HK is ATP dependent, one would expect the activity of HK to be high when respiration is increased. Lynen and Koenigsberger [70] suggested that the slowing-down of the HK-reaction might be due to feedback-inhibition of glucose phosphorylation by glucose 6-phosphate or to a decrease in ATP-concentration in the cytosol because it is compartmentalized within the mitochondria. In this case, the HK-reaction would be impaired due to lack of ATP at the site of glucose phosphorylation [69].

In addition to the above regulators of glycolysis it became clear that other effectors and mechanisms of regulation must exist. For instance, Chance [17] showed that in isolated mitochondria respiration and oxidative phosphorylation is regulated by the concentration of ADP. Shortly thereafter the concept of allosteric regulation was established [92, 93]. Allosteric enzymes are regulatory enzymes, which are inhibited or stimulated by end-products of the metabolic sequence (negative or positive feedback, respectively). The inhibitory or stimulating metabolite is known as the effector. The allosteric regulation of glycolysis in yeast was summarized in a model by Sols et al. [111] in 1971. According to this, regulation of glycolysis is based on negative or positive feedback regulation by PFK-1. A decrease of PFK-1 activity is observed when citrate and ATP concentrations increase under conditions where respiratory activity is high and glycolysis is reduced. In contrast, an increase in PFK-1 activity occurs when concentrations of ammonium ions, ADP and AMP are high, under conditions of restricted respiration and increased glycolysis. This model could account for the "Pasteur effect", namely the inhibition of glycolysis by respiration, and the "Crabtree" or "glucose effect", namely the inhibition of respiration if ample glucose is available.

3.1.2 Current Concepts

As opposed to earlier studies, more recent investigations used growing cells for the dynamic analysis of glucose and oxygen turnover rates [54, 55, 112]. Now, two different modes of glucose metabolism in yeast are described in relation to control of glucose uptake.

In obligate respirative yeast, such as *Trichosporon cutaneum*, where the specific glucose uptake rate q_S is strongly correlated to the specific oxygen uptake rate q_{O_2} (Table 1, Fig. 1). Thus, if the respiration rate is low, glucose uptake is low; in addition no ethanol is excreted, and if glucose is supplied in excess it accumulates extracellularily. The reason for this behavior is not clear. Obligate respirative fungi, such as *Chaetomium cellulolyticum*, exhibit the same growth characteristics (Fig. 2).

In contrast, in glucose sensitive yeast (facultative anaerobic), such as brewer's and baker's yeast, glucose uptake (q_S) is not controlled by the respiration rate (q_{O_2}). In these strains when sugar is supplied in excess, surplus glucose is excreted as ethanol (Table 1, Fig. 3).

Candida tropicalis represents an intermediate form between the above. When oxygen is not limiting and glucose is in excess, glucose uptake is controlled, therefore no ethan-

Table 1. Typical data of growth parameters of glucose sensitive and insensitive yeast strains. Defined media and carbon limitations were used in the chemostat. It is noteworthy that the highest glucose uptake rates are found in glucose sensitive strains and not in strict oxidative types. *Candida tropicalis* is an intermediate form excreting ethanol under conditions of O_2-limitation (for details s. [2, 49, 68/100, 104]). For abbreviations s. p 24

Cell types		Effectors	D_C h^{-1}	$Y_{X/S}$ —	q_{O_2} mmol g^{-1} h^{-1}	$q_{S_{max}}$ mmol g^{-1} h^{-1}
I	*Trichosporon cutaneum* (obligate aerobis)	O_2	0.4	0.50–0.6	8	$\cong 5$
II	*Candida tropicalis* (obligate aerobic)	O_2, (C_6)	0.6	>0.5	14	<20
III	*Saccharomyces cerevisiae* (aerobic: diauxic growth)	O_2, C_6	0.4	0.5	8	20
IV	*Saccharomyces cerevisiae* (anaerobic: monoauxic growth)	C_6	0.3	0.1	—	20
V	*Schizosaccharomyces pombe* (monoauxic growth)	O_2, C_6	0.35	0.3	1.5	$\cong 5$
VI	*Chaetomium cellulolyticum* (monoauxic growth)	O_2	0.29	0.51	7.62	0.6
VII	*Bowes melanoma cells* (diauxic growth)	C_6	0.02	0.15	0.24 [h^{-1}]	0.15 [h^{-1}]

ol is produced. In contrast, under conditions of oxygen-limitation, *C. tropicalis* excretes ethanol (Table 1, Fig. 4).

In *Schizosaccharomyces pombe* glucose uptake is not controlled. Therefore it forms ethanol. In this particular yeast ethanol cannot be used for synthesis of biomass during

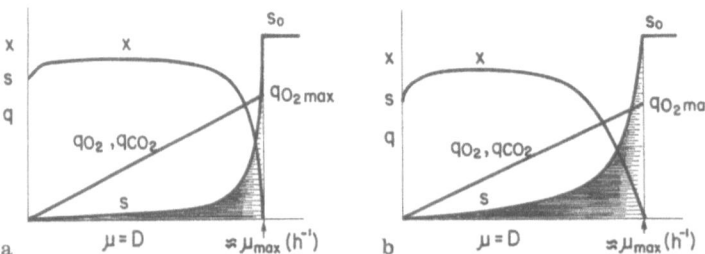

Fig. 1 a and b. *Trichosporon cutaneum* is a glucose insensitive yeast strain that is not affected by glucose or oxygen in continuous cultivation (**a**, glucose limitation; **b**, oxygen or trace element limitation; [49, 50, 59, 60]). Under the condition of oxygen or trace element limitation (Fe or Mo; **b**), the maximum specific oxygen uptake rate begins to decrease at lower dilution rates as compared to carbon limitation (**a**).

In the case of oxygen or trace element limitation D_c is reduced by about 10%. For typical values s. Table 1; for abbreviations s. p 24. The chemostat diagrams indicate that a strong glucose uptake control exists, which is strictly correlated to the respiration. As a consequence, residual glucose appears in the medium at lower dilution rates (represented by the dashed area under the S-graph) as compared to baker's yeast (s. Fig. 3). The (hitherto unknown) control mechanisms of glucose uptake by respiration prevents an overflow reaction at the point of pyruvate and no ethanol excretion appears under any condition. The activity of the glycolytic key enzymes, which are hexokinase, phosphofructo 1-kinase and pyruvate kinase, may be controlled by effectors such as ATP, fructose 1,6-biphosphate and citrate (s. also Fig. 13)

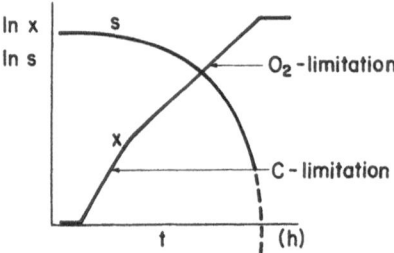

Fig. 2. Batch growth of *Chaetomium cellulolyticum* (obligate aerobic) exhibits a strict monoauxic growth pattern in batch cultures that is familiar among molds [125]. It is a glucose (and oxygen) insensitive cell type. The specific growth rate μ reaches its maximum during a first growth period when growth is carbon limited. The growth rate decreases when oxygen becomes the limiting factor, however, no ethanol is excreted. This behavior is typical for many strict aerobic fungi, where glucose uptake seems to be controlled by the respiration rate. Maximal growth rate μ_{max} of *Chaetomium cellulolyticum* is rather small, as compared to yeasts, but yield $Y_{X/S}$ and RQ similar. For typical values s. Table 1; for abbreviations s. p 24. It is noteworthy that no reliable data from chemostat experiments are available — this is due to experimental difficulties. Glucose uptake of this type of fungi is similar to *Trichosporon*, i.e. it is strongly associated with the respiratory capacity. Therefore, an overflow reaction and excretion of metabolites (ethanol, lactate, acid) does not occur

a second growth phase in batch cultures since it lacks gluconeogenetic capability (missing glyoxylic pathway). Nevertheless, ethanol is oxidized completely after glucose consumption resulting in a secondary monoauxie in batch cultivation (Table 1, Fig. 5).

Studies of *Saccharomyces cerevisiae* in continuous cultures showed that the specific oxygen uptake rate q_{O_2} is proportional to the dilution rate up to the point D_R when ethanol production begins and the oxygen uptake rate remains constant [100]. Thus, at high dilution rates above D_R a dramatic shift from a purely oxidative glucose breakdown to a mixed form of glucose degradation was noted. This observation was the starting point for the formulation of a new regulatory concept [112], in which ethanol

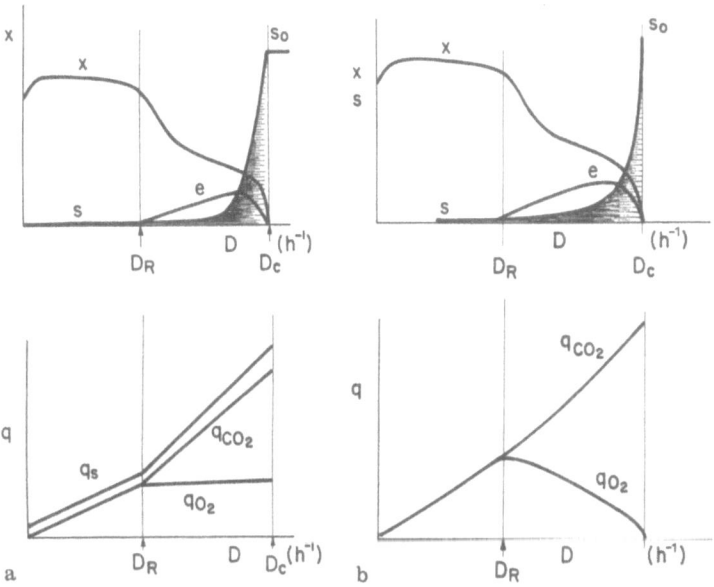

Fig. 3a and b. The growth behavior of *Saccharomyces cerevisiae* is typical for the so called glucose sensitive yeasts, in which glucose uptake is not controlled by the respiration rate. In glucose sensitive yeast strains surplus glucose is excreted as ethanol. Some bacteria, such as *Escherichia coli* exhibit a similar growth pattern leading to acetate excretion [78, 98]. Similarly, in transformed animal and human cell lines glucose uptake is increased as compared to normal cells, leading to lactate excretion [65, 66]. a Chemostat culture of *S. cerevisiae*; under the condition of carbon limitation. The specific oxygen uptake rate q_{O_2} is proportional to the dilution rate up to the point D_R, when ethanol excretion begins and the oxygen uptake remains constant until washout. b When oxygen or trace elements are limiting, D_R is markedly reduced. For typical values s. Table 1; for abbreviations s. p 24. The dashed area under the S-graph indicates that no residual glucose is present in the medium when the dilution rate is below D_R. In other words uptake is not under control of the respiration rate. This results in the excretion of ethanol since the glycolytic flux exceeds the capacity of the respiratory pathway. The area under the S-graph in b is larger than the corresponding area in a — this is due to restricted respiration. In this case the respiration is restricted by a limitation in O_2, Fe, Mo, etc. In relation to this, compare also 1a and 1b. A new regulatory concept was described recently, which explains ethanol formation by glucose sensitive yeasts as an overflow reaction of the glycolytic pathway when the respiratory capacity is exceeded [54, 55, 112]

formation is explained as an overflow reaction of the glycolytic pathway when the capacity of the TCA-cycle and mitochondrial respiration is exceeded. A schematic representation of the model and its variables is given in Fig. 6. The model is flexible enough to meet all variants of genera- and species-specific regulatory concepts of glucose breakdown.

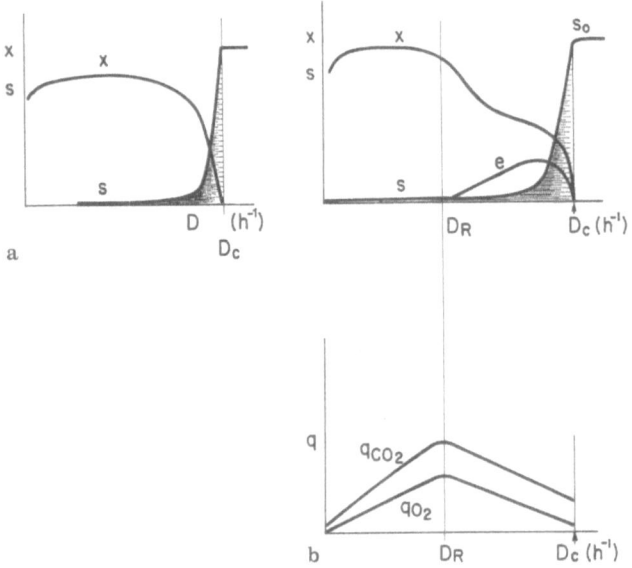

Fig. 4a and b. In *Candida tropicalis* glucose uptake and breakdown is controlled by the rate of oxygen uptake under the condition of carbon limitation (**a**). Therefore, no ethanol is produced (monoauxic growth in batch culture). In contrast, under conditions of oxygen limitation (**b**) this yeast excretes some ethanol (diauxic growth in batch culture). In continuous cultivation *Candida tropicalis* shows similar growth characteristics as *Trichosporon* and related fungi when growth is carbon-limited. Because *C. tropicalis* has a high respiratory capacity, glucose is completely oxidized until washout. The RQ remains constant at all dilution rates. For typical values s. Table 1; for abbreviations s. p 24. However, under conditions of oxygen limitation, ethanol is produced above a dilution rate $D_R \cong 0.3$ (which depends on the degree of limitation) that is similar to growth characteristics of *Saccharomyces*-type yeast (s. Fig. 3)

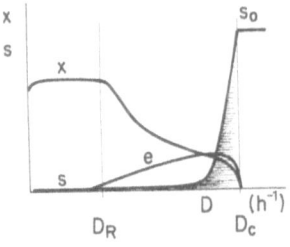

Fig. 5. In the fission yeast *Schizosaccharomyces pombe* glucose uptake is not controlled by respiration, thus ethanol is excreted (carbon-limitation). In addition, it is very likely that this yeast is also oxygen sensitive. *S. pombe* has an unusual low yield both in batch and continuous cultivation, and maximum growth rates (μ_{max}) are rather small. For typical values s. Table 1; for abbreviations and symbols s. p 24

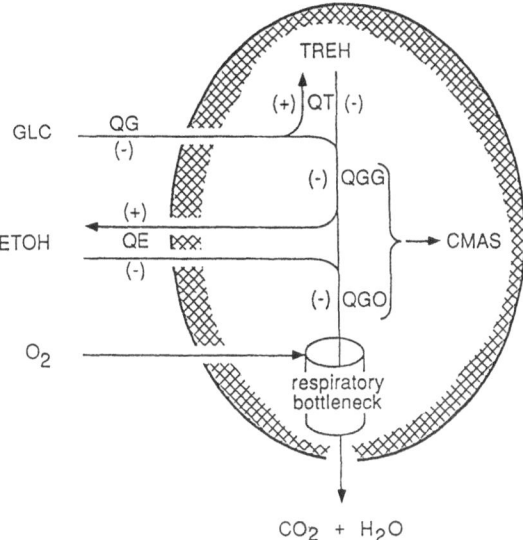

Fig. 6. Schematic representation of the "bottleneck model" and its variables [116, 117]. This model is able to describe events observed both in long-term metabolic control (steady-state conditions in chemostat culture) as well as in short-term metabolic fluctuations during the cell-cycle (s. Figs. 9, 10). Generally, sources and sinks are assigned as positive and negative fluxes, respectively. CMAS, residual biomass; ETOH, ethanol; GLC, glucose; TREH, storage materials; Q, specific fluxes; QG, specific glucose uptake rate; QE specific ethanol uptake/excretion rate; QGG, glycolytic flux, QT, rate of storage material formation/consumption; QGO, glucose flux through TCA-cycle and mitochondrial respiratory chain

3.2 Tumor Cells

With regard to glucose metabolism, tumor cells and baker's yeast are similar. This similarity of glucose metabolism was first observed more than 60 years ago using resting cells. Glucose molecules enter tumor cells via facilitated diffusion, and are phosphorylated immediately by hexokinase (HK) in the cytoplasm (s. Sect. 7). The HK-reaction occurs with high velocity, even at low glucose concentration. Thus, glucose cannot be detected inside the cell in its unphosphorylated form. After phosphorylation, glucose 6-phosphate enters glycolysis and, eventually, the TCA-cycle. A difference between yeast and animal glucose metabolism is that animal cells excrete lactate as a by-product of glycolysis whereas baker's and related yeasts excrete ethanol. The pathways for ethanol and lactate production involve the same enzyme catalyzed pathways with the exception that *Saccharomyces*-type yeast decarboxylate pyruvate, rather than reducing it directly as in animal cells, giving acetaldehyde, which is then reduced to ethanol.

Although tumor cells are more difficult to cultivate than yeast, data obtained from tumor cells grown in chemostat systems is available (Fig. 7). In Bowes melanoma cells the specific glucose uptake rate q_s and specific lactate production rate q_p both increase proportionally with increasing dilution rate [65]. It is important to note that the diagram shown in Fig. 7 does not extend over the whole range of D. Especially very low dilution rates below $0.15 \, d^{-1}$ have not yet been investigated. This is due to a technical obstruction, i.e. the extremely low pump rates at low dilution rates, which has not yet been resolved. In relation to glucose sensitive yeasts, it is not known whether an equivalent point of D_R exists for animal cells. The production of lactate is tumor cell specific but can occur in normal cells too.

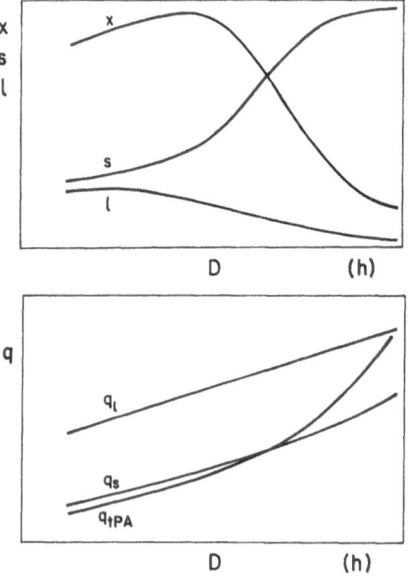

Fig. 7. Bowes melanoma cells are a human tumor cell line which excretes lactate when grown on glucose as a carbon source [65, 66]. The cell line presented here can grow in suspended culture in a well defined, serum- and protein-free synthetic medium (31 defined components). It was obtained by careful selection and it is stable. In continuous cultivation growth characteristics resemble those of *S. cereyisiae*, which is a glucose sensitive yeast (Fig. 3). Specific lactate production rate q_p and specific glucose uptake rate q_s both increase proportionally with increasing growth rate ($D \equiv \mu$). For typical values s. Table 1; for abbreviations s. p 24. In batch culture lactate is formed during a first growth stage, which can be metabolized subsequently in a secondary growth phase (diauxic growth). It is noteworthy that the specific rate of t-PA formation (tissue type plasminogen activator) is strictly growth associated. This is illustrated by the q_p graph in the lower panel, which increases proportionally with increasing growth rate

The study of tumor cells may provide information on what principles govern glucose degradation and lactate formation in these cells. However, not much is known about how the original cells become transformed. Nevertheless, a number of changes are known to take place. These include:

— inappropriate expression of proto-oncogenes or oncogenes [23, 27, 110];
— alterations in cell cycle regulating processes [27, 119];
— changes in morphology and/or loss of organization and anchorage-dependence [23];
— changes in cytoskeleton and adhesion plaques [23];
— alterations in intercellular communication by alteration of gap junctions [23, 124];
— changes of cell surface signal transduction, i.e. polyphosphoinositide system and second-messenger systems [9, 73, 83];
— release of transforming growth factors and proteases [23, 27];
— reduction of mitochondrial number [87];
— increase in aerobic glycolytic activity [23, 29, 92, 93].

It is interesting to note that just one change in proto-oncogene expression leads to increased glucose uptake and lactate excretion.

4 Proto-Oncogenes and Oncogenes

Today more than 30 proto-oncogenes have been identified that may become oncogenes after appropriate alteration or insertion [11, 56, 132]. These genes are very conservative i.e. extended regions have remained constant during the evolutionary process. For instance, the *scr* gene has been found in both vertebrates and insects, and homologous regions of the *ras*, *has*, and *bas* proto-oncogenes were identified in human cells as well as in yeast cells [24, 31, 35].

The proto-oncogenes have a very broad scope of action [11], including control of transcription, cell cycle, signal transduction, and the regulation of gene expression. Thus, the activation of some proto-oncogenes plays a crucial role in control of cell growth and carcinogenesis. Some of the triggers and modulators controlling cell proliferation in both normal and transformed cells consist of a system designed to link the genes to their environment. This system is a group of membrane receptors and regulatory units, which convert and amplify external signals to produce second messengers that alter gene expression. Changed perception may activate proto-oncogenes and provoke cell transformation.

In contrast to transformed cells, the normal expression of certain proto-oncogenes is obviously well timed and takes place in temporal order during the cell cycle [5, 27]. The gene products of oncogenes are [11]:

— protein kinases that phosphorylate amino acid residues of other proteins (e.g. *scr*);
— p21 proteins with guanine nucleotide-binding properties (e.g. the *ras*-family);
— nuclear proteins possibly involved in regulation of gene expression (e.g. *myc*, *myb*, *fos*, *jun*); and
— growth factors that are secreted by the cell (e.g. *sis*).

During the past years a lot of evidence has been obtained that some cellular homologues (c-*onc*) of the viral transforming genes (v-*onc*) are crucial for the regulation of normal cellular development and differentiation in eukaryotic cells. Furthermore, expression of these homologues, e.g. the human c-H-*ras*, can lead to neoplasia in transgenic mice [91]. Similarly, deregulated expression of another cellular oncogene c-*fos* expression interferes with normal bone development in transgenic mice [101]. This supports the hypothesis that the *fos* gene "may be central to the cell's ability to convert short-term stimulation to long term responses — such as growth and memory formation" [74, 123].

5 Eukaryotic Cell Cycle

The behavior of any living cell is strongly linked to the cell cycle and metabolic phenomena observed have to be correlated to this basic event. Eukaryotic cells undergo the cell cycle in order to grow and multiply (Fig. 8). It has been shown that the control and coordination of this cycle involves many proto-oncogenes [5, 27, 74].

Tumor formation is usually observed in tissues where the individual cell retains the capability to proliferate (e.g. liver, kidney, glands, connective tissue, etc.). These cells may re-enter the cell cycle from the resting phase and proliferate after the appropriate stimulus. Alternatively, cells in in a process of continuous proliferation (e.g. epidermis, epithelia, blood stem cells) may never enter the resting phase at all (Fig. 8). In contrast, tumors are rarely seen in tissues with no or poor possibility of regeneration (e.g. nerve and striated skeletal muscle tissue). These cells usually remain in the resting phase after completion of cell differentiation.

Animal cell

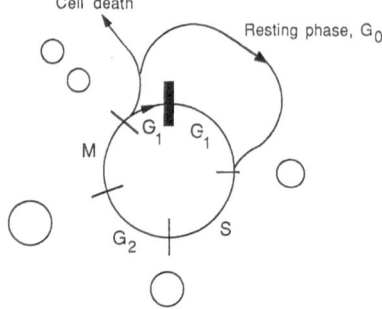

Schizosaccharomyces pombe

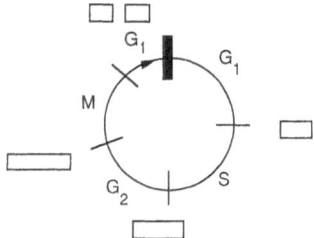

Saccharomyces cerevisiae

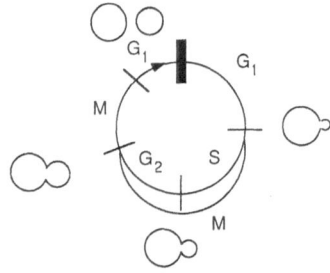

Fig. 8. Schematic representation of animal cell cycle (top), fission yeast cell cycle (middle) and budding yeast cell cycle (bottom). The cycle runs clockwise. Incidence of cell death and resting phase are indicated in the animal cell cycle only, but they are similar in all eukaryotic organisms. It is of interest that the basic pattern is universal in all eukaryotic cells. However, in budding yeast there is no real G_2-phase. A short mitotic spindle is formed during the S-phase and the cell becomes arrested in midmitosis for the rest of the cell cycle [45]. Homologues of the cell division cycle control genes have been found in yeast and animal cells (for details s. Sect. 5.). For instance, the *cdc2* gene controls the cycle at the corresponding key points in all 3 organisms as indicated by black bar. G_1, gap 1; S, DNA-synthesis; G_2, gap 2; M, mitosis. Adapted from Murray [81]

Today, evidence suggests that the control of cell cycle and length of the resting phase is strongly influenced by a group of genes called cell division cycle control genes (*cdc*-genes) including some proto-oncogenes [5, 27]. Cell cycle and growth regulating genes have been highly conserved in evolution. Homologues of these genes, e.g. *cdc2* and *cdc10*, have been found in budding and fission yeast, animal and human cells [14, 64, 84]. This implies that these genes may serve the same function in many types of eukaryotic cells [81]. We will now discuss the control of the cell cycle in one of these examples.

The cell division cycle of *Saccharomyces cerevisiae* synchronizes spontaneously during respirative (oxidative) glucose breakdown (no ethanol present in the medium) in continuous culture (Fig. 9) [103, 149]. Several investigations of enzyme expression in yeasts during the cell division cycle showed that various enzymes undergo remarkable changes during the cell cycle [30]. Our own studies on *Saccharomyces cerevisiae* revealed that activities of a variety of enzymes (Table 2) fluctuate in synchrony with the cell cycle of growing yeasts [2, 3, 53, 67]. In contrast, levels of mitochondrial cytochrome contents remained relatively unchanged in *S. uvarum* throughout the cell cycle. Cytochrome contents were related more to long-term adaptations in relation to the metabolic status [39, 88]. In synchronized cultures, ethanol production takes place but fluctuates during the yeast cell cycle [53]. Low quantities are produced at the beginning of the division cycle that act as an inducer for gluconeogenic enzymes since ethanol can be used as a carbon source [3].

More recently, the model of Sonnleitner and Käppeli [54, 55, 112] was extended (Fig. 10) to include the process of synchronization in order to understand fluctuations of ethanol concentrations, gas exchange rates (specific carbon dioxide production rate, q_{CO_2}, s. Fig. 9), and enzyme levels during the cell cycle [116, 117].

As yet, it is not known whether fluctuations in gas exchange rates and enzyme activities occur throughout the cell cycle in animal cells as they do in yeast cells. Also, it is not clear if lactate or another metabolite can exert a similar repression of enzymes in tumor cells as high concentrations of ethanol do in yeast cells. In relation to this, it has been shown that accumulating lactate did not affect growth and antibody production in murine hybridoma cells even at high lactate concentrations of 2.5 g l^{-1}, if adequate pH control was provided [99].

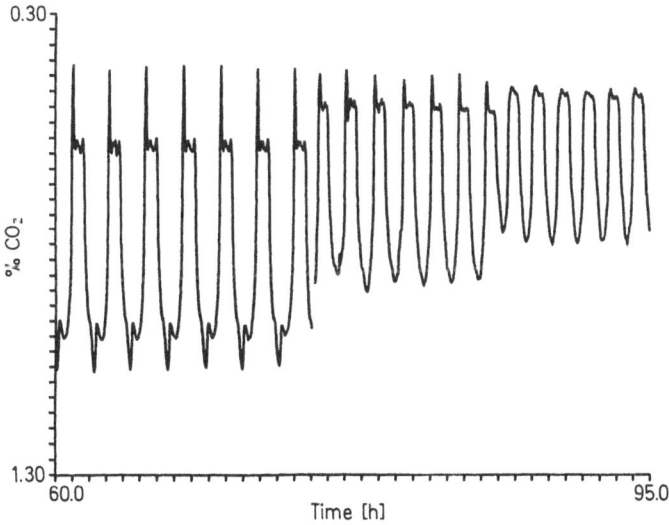

Fig. 9. Stable synchronization of growth in a chemostat experiment (*Saccharomyces cerevisiae*). Essentially, dilution rates of D = 0.182, 0.213 and 0.222 were set in a highly accurate chemostat [75]. The dominant peaks of CO_2-release refer to the initiation of a new cell cycle of the synchronized (parental/filial) subpopulations. Parameter of initiation is the availability of (extra- plus intracellular) glucose. Trehalose is activated as carbon source with a slight overshoot of CO_2 and $NADH_2$ [77, 61–63]. In contrast to this system, synchronized growth is not found in *Trichosporon cutaneum* and *Candida tropicalis* cultures

Table 2. Regulation of enzymes and cytochromes obtained from chemostat experiments. The cells were harvested during steady state conditions and cell cycle stages as indicated. $<D_R$, dilution rates where purely oxidative glucose breakdown noticed; $>D_R$, dilution rates where repressed or oxido-reductive glucose metabolism was observed. Stable synchronizations are observed only if the cells are grown at $D < D_R$, i.e. when glucose is completely respired and no ethanol does accumulate in the medium. For abbreviations s. p 24
Steady state conditions

Enzymes [Unit]	Strain	$<D_R$	$>D_R$	Refs.
ATPase, membrane bound [mmol g^{-1} h^{-1}]	S. cerevisiae	6	$\cong 16$	[67]
Carnitinacetyl transferase [mmol g^{-1} h^{-1}]	S. cerevisiae	240	$\cong 0$	[18]
Malate dehydrogenase [mmol g^{-1} min^{-1}]	S. uvarum	5–6	$\cong 1$	[39]
Cytochrome a [nmol g^{-1}]	S. uvarum	30–50	$\cong 25$	[88]
Cytochrome b [nmol g^{-1}]	S. uvarum	50–100	$\cong 50$	[88]
Cytochrome c [nmol g^{-1}]	S. uvarum	150–320	120–150	[88]

Dependence from cell cycle ($D < D_R$; *Saccharomyces cerevisiae*)

Enzyme [Unit]	Single cell phase[a]	Double cell phase[b]	Refs.
Fructose-diphosphatase [mmol min^{-1} g^{-1}]	<1	11	[3]
Phosphoenolpyruvate-carboxy kinase [mmol min^{-1} g^{-1}]	12	23	[3]
Carnitinacetyl transferase [mmol g^{-1} h^{-1}]	19	26	[3]
Malate dehydrogenase [mmol min^{-1} g^{-1}]	3.5	5	[3]
ATPase, membrane bound [mmol g^{-1} h^{-1}]	$\cong 2.2$	$\cong 1.3$	[67]
Cytochrome a [nmol g^{-1}]	>30	<30	[2]
Cytochrome b [nmol g^{-1}]	≥ 40	<40	[2]
Cytochrome c [nmol g^{-1}]	>100	80	[2]

[a] Single cell phases: corresponds to late M and G_1;
[b] double cell phases: corresponds to S and G_2 phase (s. Fig. 8).

From a metabolic standpoint of view, growing yeast cells share much in common with normal and transformed animal cells. This is illustrated by the fact that glucose insensitive yeast (*Trichosporon cutaneum* and *Candida tropicalis*) and normal untransformed cells do not excrete ethanol and lactate, respectively. In contrast, glucose sensitive yeast and tumor cells do excrete ethanol and lactate, respectively. Moreover, yeast and animal cells show many similarities in their regulation of the cell cycle, transcription and growth. For instance, yeast and animal cells share *cdc2* and *cdc10* [14, 64, 81]. The gene product of *cdc2* is a protein kinase which possibly stands at the apex of a protein cascade (s. Sect. 6.) thus potentially controlling cell cycle at the molecular level [45]. Furthermore, both cell types possess similar transcription activator proteins since *Saccharomyces cerevisiae* (*GCN4*) and the DNA-binding *jun* onco-protein from a chicken sarcoma show significant homology [115]. It is now believed that *jun* and *GCN4* are equivalents to the mammalian protein AP-1 which stimulates transcription [13, 106, 115, 122]. Moreover, striking similarities were identified among the *ras*-, *bas*-, and *has*-related proteins of yeast, insects, molluscs, and mammals [5, 24, 31, 35, 41].

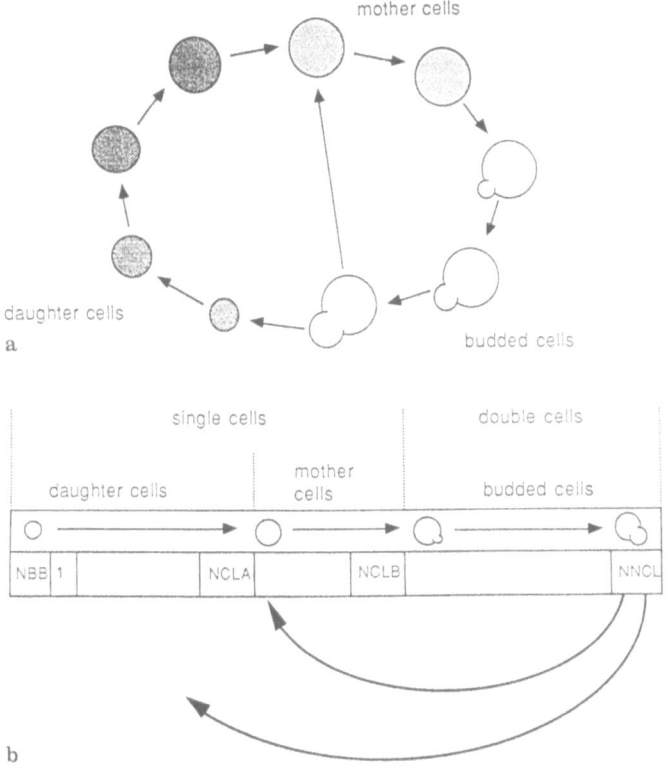

Fig. 10a and b. Schematic diagram of **a** the cell cycle of Saccharomyces cerevisiae and **b** its translation into a 1-dimensional array of cell classes. A model based on these assumptions was developed [116, 117] in order to understand the oscillations of short-term growth kinetics and enzyme activities related to the cell cycle more thoroughly. The model does not distinguish between mother cells and grown-up daughter cells. NBB, baby cell class; NCLA, cell class A; NCLB, cell class B; NNCL, last cell class

One major difference between yeast cells and normal untransformed cells is that yeast cells start to divide as soon as they meet adequate growth conditions. However, the cell cycle of normal cells is more strictly controlled by the growth regulating genes. In animal cells a more specific trigger of cell division is required in addition to adequate growth conditions. It is interesting that carcinogenesis in yeast has not been noticed, although retrovirus-like particles have been identified in yeast [76, 82].

6 Cell Surface Signal Transduction

It is clear that some oncogenes are involved in the cellular signal transducing pathways (Fig. 11). Changes in signal transduction can result in massive changes of cell behavior and may lead to cell transformation. Evidence suggests that the characteristic increase of glucose uptake during cell transformation could be related to changes in signal transduction. Especially since products of signal transduction, such as second messengers and activated protein kinases influence cell division and other functions [6, 8, 9, 36, 45, 73, 83]. Thus, insight into the molecular events involved in signal transduction may lead to a clearer understanding of the process of carcinogenesis.

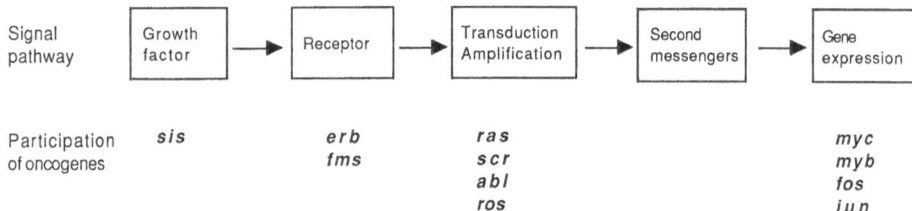

Fig. 11. Major components of signal transduction pathways. The position where distinct (proto-) oncogenes may participate in signaling is depicted. The individual oncogenes, as indicated, are thought to be involved in the following structures or sequences: *sis*, growth factor that shares much in common with the platelet derived growth factor (PDGF); *erb*, portion of epidermal growth factor (EGF) receptor; *fms, scr, ros* and *abl*, membrane associated proteins (e.g. protein kinase); *ras*, G protein of adenylate cyclase; *myc, myb, fos* and *jun*, proteins that are localized in the nucleus and involved in growth control and cell differentiation control. It is noteworthy that the control of growth and differentiation in yeast may be governed by genes that are homologous to the animal genes. For instance, it was found that the GCN4 gene of *S. cerevisiae* encodes for the same product as *jun* (for details s. text). Adapted from Berridge [7]

The steps of signal transduction are outlined in Fig. 12. Briefly, signal transduction is comprised of two pathways, which may be interrelated, namely the phosphoinositol system and the cyclic adenosine monophosphate system (cAMP system). When hormones, neurotransmitters and growth factors bind to receptors a transmembrane signal is generated [107]. It is transmitted via GTP-binding proteins (Gp, [37]) to phosphodiesterase or adenylate cyclase. Second messengers are then generated, such as inositol triphosphate (IP_3), diacylglycerol (DAG), Ca^{2+}, and cAMP [9, 36, 57, 73].

The cellular response to these second messengers include the following: oncogene expression [9]; formation of growth factors, e.g. prostaglandins [95, 96]; activation of a $Na+$-$H+$ exchanger in the cell membrane [9]; activation of DNA synthesis, and activation of protein kinase cascades [113, 114].

One of the unifying themes of biochemical regulation is that the phosphorylation of proteins by protein kinases and their dephosphorylation by phosphoprotein phosphatases lead to changes in cell function [48, 58]. Protein kinases modify the activity of other proteins by phosphorylation of amino acid residues. Membrane associated protein kinase C that is Ca^{2+} and phospholipid-dependent is greatly activated by DAG (Fig. 12; [6, 57, 85]). Tumor-promoting phorbol esters may substitute for DAG by permanently stimulating protein kinase C [83]. Activated protein kinase C enhances the catalytic function of its target enzymes by phosphorylation of serine and threonine residues. In relation to carcinogenesis and alterations in glucose metabolism, the effects of protein kinase C include the phosphorylation of the glucose transporter in human erythrocytes [136] and the activation of the *fos, myc* and *myb* oncogenes [7, 9, 73].

In animal cells changes in signal transduction have also been linked to the expression of other oncogenes [33, 56, 89, 126]. For instance, it was shown recently that the *ras*

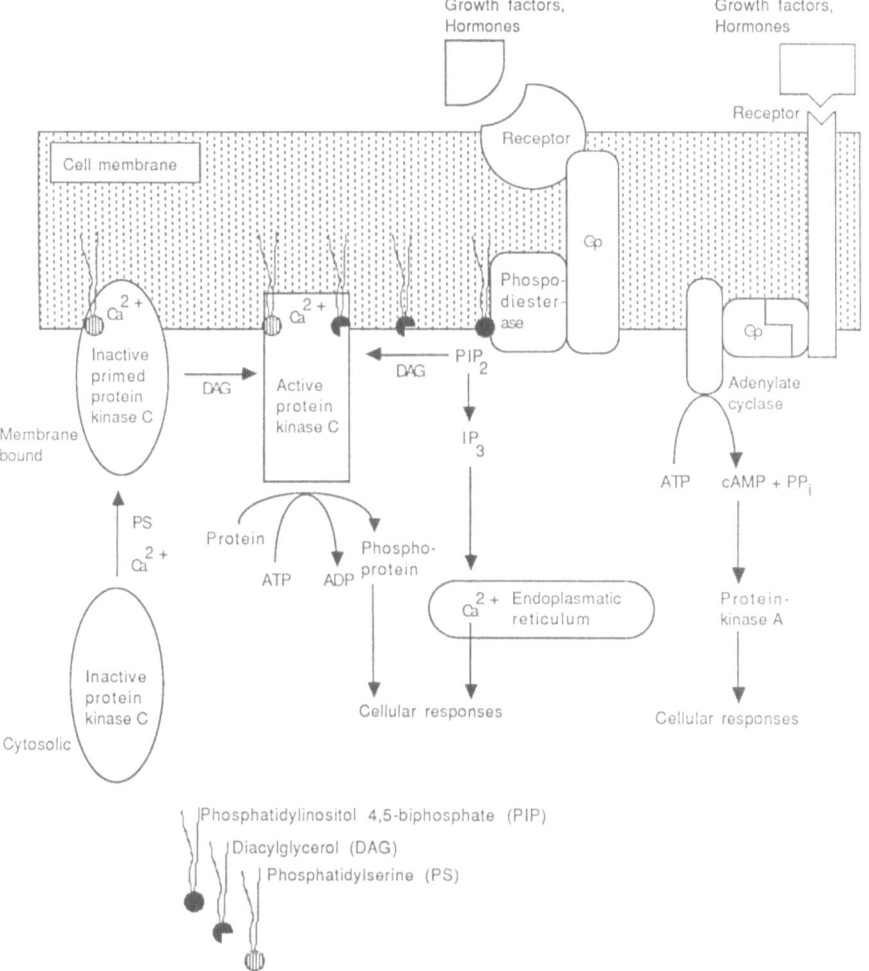

Fig. 12. A simplified view of the phosphoinositol system (left side of the panel) and the cAMP system (right side of the panel) is depicted. Hormones and growth factors bind to the specific receptors. The messenger systems are activated via two types of receptors, stimulating and inhibitory, which are linked to phosphodiesterase or adenylate cyclase by specific regulatory proteins (Gp, GTP-binding proteins). For convenience, only the stimulatory Gp of the signal transduction are shown. The activated phosphodiesterase splits phosphoinositol 4,5-biphosphate (PIP$_2$) to diacylglycerol (DAG) and ino-sitoltriphosphate (IP$_3$). IP$_3$ leads to Ca^{2+} release from the endoplasmatic reticulum. Ca^{2+} and DAG activate proteinkinase C which subsequently phosphorylates target proteins. Likewise, activated adenylate cyclase splits ATP to cAMP and PP$_i$. cAMP can activate the cAMP dependent protein-kinase cascade. In this simplified scheme the role of the cAMP messenger system in regulating the responsiveness of a cell to activation through the phosphoinositol system is not included (cf. [95]). Again, the signal transduction pathways of yeast and animal cells may share much in common. Lately, it was shown that nucleotide and amino acid sequences of G proteins are highly homologous among yeast and animal cells. Moreover, the function of a defective G protein in a yeast mutant was restored by inserting the rat gene into the yeast's genome (for details s. Sect. 6)

oncogene may encode for a Gp that transduces the signal generated by growth hormone receptors leading to stimulation of phosphodiesterase [126]. Alternatively, *ras* gene products might be related to structural changes in the cytoskeleton as well [41]. However, in yeast cells *ras*-like gene products were found in the cell membrane, but they seem not to be involved in signal transduction [102, 105]. The *scr* gene is another example of an oncogene which may be involved in intracellular signal transduction and the protein kinase cascade. The expression of this oncogene alone leads to cell transformation [40]. The gene product of *scr* is a membrane associated protein kinase which comprises a phosphorylated protein of 60 kDa ($pp60^{v-scr}$; [137]). It is thought that this kinase is involved in signal transduction by enhancing the activity of phosphodiesterase [118].

In relation to the regulation of glucose metabolism in animal and yeast cells, it would be of interest if similarities were found in the signal transduction mechanism of these cells. Miyajima et al. [79] characterized the nucleotide and amino acid sequences of *GPA1* protein of *Saccharomyces cerevisiae*, which is thought to be a subunit of a yeast Gp. They showed that the *GPA1* subunit is highly homologous to a subunit of mammalian Gp. Moreover, Dietzel and Kurjan [28] identified another yeast gene, *SGC1*, that also shows a high level of homology to the α subunits of mammalian Gp. In this case it was even possible to restore defective Gp function in yeast mutants by insertion of the corresponding rat gene that encodes for the mammalian G_α subunit [28]. These findings indicate that signal transduction pathways are very highly conserved among eukaryotic cells.

7 Control of Glucose Metabolism in Transformed Cells: Underlying Mechanisms

It appears that the role of some oncogenes may contribute to the regulation of glucose metabolism in tumor cells. Nevertheless, the reasons for increased glucose consumption and lactate production after cell transformation are not known. Speculations on the underlying mechanism for this concentrate presently on:

— factors affecting the TCA and/or respiratory chain, and,
— increased activity of the glycolytic pathway and/or permissive glucose uptake.

7.1 Respiration

Warburg's suggestion that the respiration is impaired in tumor cells led to extended controversies about the origin of cancer [15, 87, 92, 93, 127, 128, 130, 133]. His idea was abandoned later because oxygen consumption was found to be only slightly reduced in tumor tissues in comparison to the original normal tissue [16, 22, 87, 133]. More recently, this idea was taken more seriously.

A number of hypotheses have been proposed to explain the slight reduction of respiration following cell transformation [29, 87], such as:

— "reductive repression" of mitochondrial function,
— defective NADH-shuttles through the mitochondrial membrane,

— decreased rate of pyruvate oxidation, and,
— reduced number of mitochondria.

Generally, limitations in aerobic capacity lead to a decrease of intracellular ATP levels. The inhibitory action of ATP on PFK-1 is decreased (Figs. 13, 14). Increases in the PFK-1 activities affect the other two key enzymes. Hexokinase (HK) is stimulated by lower glucose 6-P levels and pyruvatekinase (PK) is stimulated by higher fructose 1,6-P levels and lower ATP levels.

Reductive Repression

This notion refers to the phenotype of cells with exhausted respiratory capacity. It is observed in respiration impaired *E. coli* [47] and in yeast cells [3, 32, 55]. However, no information is available about this mechanism in animal cells.

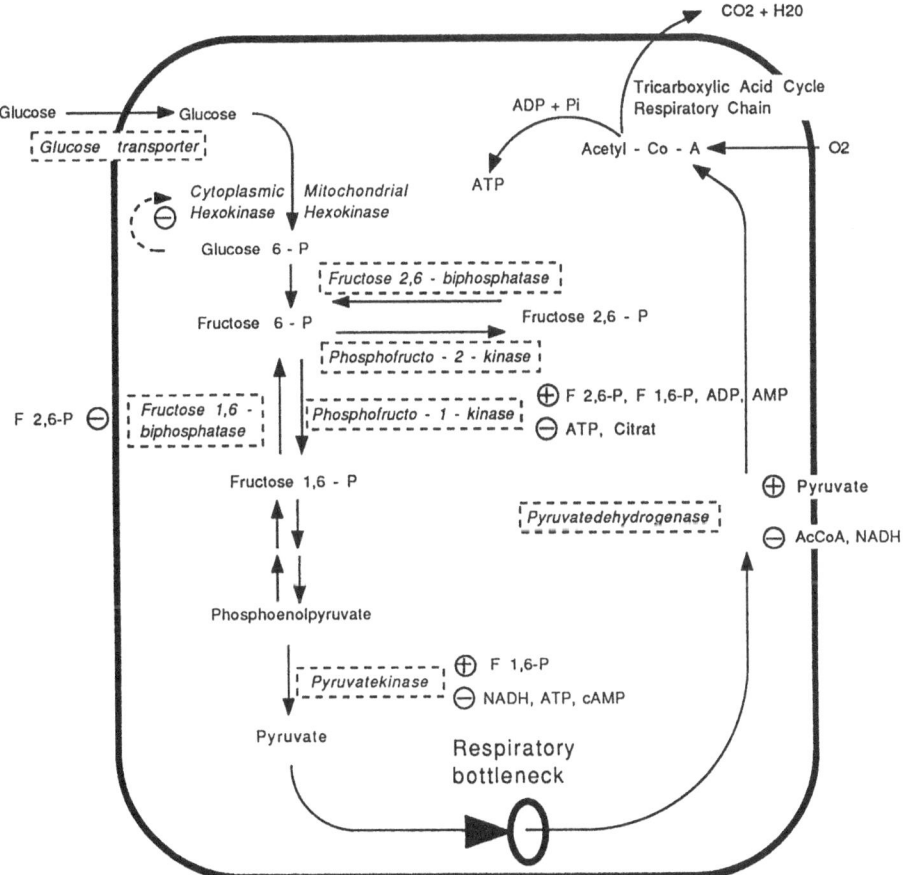

Fig. 13. Normal cell. Regulation of glycolysis in normal animal cells. The glucose flux through this pathway is restricted by ATP and citrate which inhibit PFK-1. Low levels of F 1,6-P are not sufficient to offset ATP and citrate inhibition of PFK-1. Likewise, F 2,6-P levels are also not able to activate this enzyme. High levels of G 6-P inhibit the HK reaction. PDH is fully activated by pyruvate. The respirative capacity, indicated as respirative bottleneck of the cell, is sufficient to oxidize glucose completely. -------- Enzymes that are regulated by proteinkinases and proteinphosphatases

Defective NADH-shuttles

Regarding this mechanism, it has been shown by several authors (cf. [87]) that the mitochondria operate at a sufficient rate even in those from highly glycolytic tumor cells. Therefore, this does not explain the slightly reduced respiration in tumor cells.

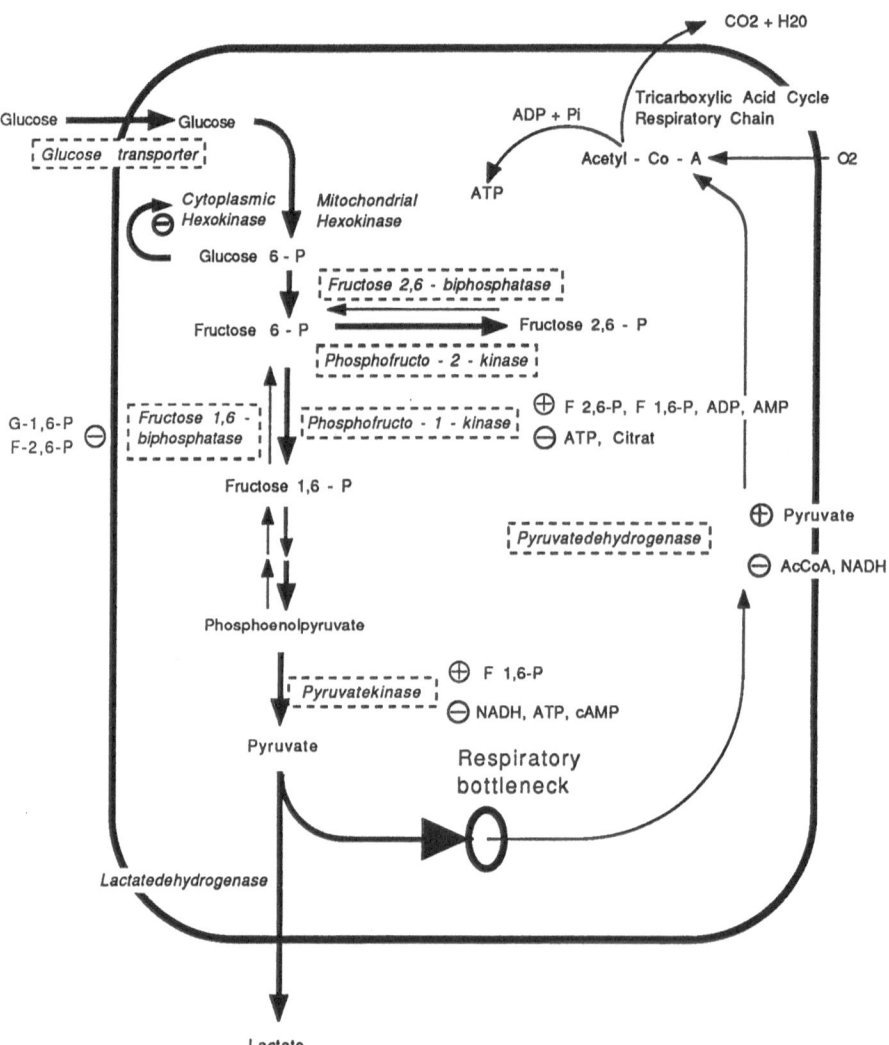

Fig. 14. Tumor cell. Regulation of glycolysis in tumor cell. Glucose uptake is stimulated markedly by enhanced synthesis of glucose transporters. The activity of mitochondrial HK is several times higher than the cytoplasmic HK activity. High F 2,6-P and ADP levels stimulate PFK-1 activity. Furthermore, high F 1,6-P levels enhance activities of both PFK-1 and PK. Accumulating pyruvate cannot be removed by PDH, because of the limited respirative capacity, indicated as respirative bottleneck. Therefore, the accumulated pyruvate has to be reduced to lactate in order to re-oxidize NADH. -------- Enzymes that are regulated by proteinkinases and proteinphosphatases

Rate of Pyruvate Oxidation

One of the first rate controlling steps of the TCA-cycle and respiration is the oxidation of pyruvate by pyruvate dehydrogenase (PDH). Decreased PDH activity (Figs. 13, 14) leads to an accumulation of pyruvate which in turn leads to the production of lactate. Part of the PDH multienzyme complex is a protein kinase which can phosphorylate the enzyme at a serine residue. This phosphorylation of PDH inhibits its activity leading to the accumulation of pyruvate. The catalytic function is enhanced after dephosphorylation of the enzyme by a phosphoprotein phosphatase [58, 94]. This kind of PDH-regulation has been found in all eukaryotic cells, including yeast [120]. However, Bissell et al. [12] showed that the glucose flux into the TCA-cycle is unchanged upon Rous sarcoma virus transformation. This suggests that pyruvate oxidation is not affected after transformation.

Number of Mitochondria

Alternatively, reduced respirative capacity of tumor cells may be partly due to a low mitochondrial content [87]. Similarly, Arreguín [2] found a lowered mitochondrial protein content in yeast growing above D_R in chemostat experiments after a prolonged time of cultivation.

7.2 Glycolysis and Glucose Transport

Regulation of Glycolysis in Tumor Cells

The second explanation for increased glycolytic activity of transformed cells is enhanced activity of rate-limiting enzymes, namely HK, PKF-1 and PK. Experiments suggest that the activity of several glycolytic enzymes, including rate-limiting steps, are affected upon transformation [29, 108, 109, 121]. It was found for example that isoenzymes of phosphoglycerate mutase, enolase, lactate dehydrogenase and PK are phosphorylated at tyrosine residues after cell transformation by Rous sarcoma virus [20, 38, 64, 90]. The gene product of the *scr* gene, a protein kinase, phosphorylates proteins at tyrosine residues. However, other enzymes (e.g. HK and PFK-1), which also regulate glycolysis, are not modified [64]. Therefore, the role of this protein kinase in phosphorylation of glycolytic enzymes is not clear [21].

Increased glycolytic activity in mammalian cells can be attributed to increased amounts of mitochondrial HK isoenzyme as pointed out by Eigenbrodt et al. [29] and Wilson [135]. This was found in all cells exerting high aerobic glycolysis. It is of interest to note that tumor cells contain a higher amount of mitochondrial HK than the corresponding untransformed cells [29, 135]. Unlike the other HK isoenzymes, mitochondrial HK is not inhibited by low concentrations of glucose 6-P.

The second key enzyme for control of aerobic glycolysis is PFK-1. When its activity is low fructose 6-phosphate accumulates, leading to the formation of fructose 2,6-biphosphate (F 2,6-P) due to the action of phosphogructo 2-kinase (PFK-2; Figs. 13, 14). PFK-1 is markedly stimulated by F 2,6-P [121]. Insulin and phorbol esters at mitogenic concentrations cause a rise in the concentration of F 2,6-P which in turn increases PFK-1 activities in chicken embryo fibroblasts [121]. Increases of F 2,6-P are also obtained with a series of agents known to interact with proteinkinase C. These presumably act by increasing the activity of PFK-2 [121].

The third key enzyme of glycolysis, PK consists of several isoenzymes, of which PK-M$_2$ is dominant in all cells with high aerobic glycolysis [29]. In relation to the products of oncogenes, it is interesting to note that PK-M$_2$ can be phosphorylated and thus inactivated by a cAMP dependent protein kinase in vivo and in vitro [29, 90]. The significance for a restriction of the activity of PK is to prevent the loss of glycolytic intermediates like glycerol 3-phosphate to guarantee serine, purine and lipid biosynthesis in the rapidly proliferating tumor cells [29].

Glucose Transport

Increased glucose consumption in tumor cells could result from a stimulation of glucose transport through the cell membrane. Most mammalian cells take up glucose by facilitated diffusion driven by the glucose gradient, which is catalyzed by a membrane protein [23, 134]. This glucose transporter has been studied in detail using red blood cell ghosts, which consist of pure cell membranes. Recently, the sequence of this transporter was determined using expression cloning of the protein in *Xenopus* frogs and subsequent cDNA sequencing [46]. It is a glycoprotein with a molecular weight of 55 kDa and comprises 5% of the erythrocyte's membrane proteins [1]. It was also found that the glucose transporter in kidney, liver and erythrocyte cells is identical [80]. However, no homology was found between the mammalian facilitated glucose transporter, the Na$^+$/glucose co-transporter (abundant in kidney and intestine cells), and the bacterial sugar transport proteins [46]. Furthermore, it was shown that the facilitated glucose transporter is specific for D-glucose, although structurally related monosaccharides (D-mannose, D-galactose) are transported at a slower rate at a given concentration [23]. The K_m value for D-glucose is 1.6 mM as opposed to 3 M for L-glucose [52].

Following the findings of Hatanaka et al. [42-44] it became clear that the expression of v-scr (pp60^{v-scr}) results in increased glucose transport in cells infected with Rous sarcoma virus. It was suggested that alterations in the glucose transport system could play a key role in the process of cell transformation. Theoretically, the expression of the *scr* oncogene could:

a) increase the number of glucose transporter molecules in the membrane,
b) increase the activity of existing glucose transporters in the membrane, or
c) induce the synthesis of a new class of transporters.

The results of Salter et al. [103] strongly support the first hypothesis. They were able to show that expression of *scr* oncogene increases the rate of glucose transport by increasing the number of transporters in the membrane. Furthermore, they showed that antibodies raised against the purified human erythrocyte glucose transporter specifically precipitated a similar protein in the membrane fraction of Rous sarcoma virus-transformed chicken cells. The increased amount of this protein after malignant transformation was directly proportional to the increase in glucose transport rate [103]. In addition, in normal erythrocytes phosphorylation of the glucose transporter by protein kinase C enhances the uptake of glucose [136]. This observation explains the ability of phorbol esters to stimulate glucose uptake by increasing phosphoinositol turnover and protein kinase C activity [103]. Flier et al. [34] have obtained direct evidence for an increase of glucose transporter synthesis after transfection of mammalian cells with *ras* and *scr* oncogenes and after phorbol ester stimulation. They measured increases of transporter protein and glucose transporter messenger RNA in the cell membrane of transfected rodent fibroblasts. In contrast, in *myc*-transfected cells, the glucose uptake rate did not change. At the same time, Birnbaum et al. [10] used rat fibroblasts transformed with either a wild-type or a temperature-sensitive Fujinami sarcoma virus to study the underlying mechanism of glucose transporter synthesis. Within 30 min following the shift of the temperature-sensitive RSV-transformed cells to the permissive temperature, an increase of glucose transporter gene transcription and an increase of transporter mRNA was noted. This gave a 5 to 10 fold increase in total cellular glucose transporter protein.

8 Conclusion

Control mechanisms for glucose degradation in eukaryotic cells have been in the focus of interest for more than 100 years. In both glucose-sensitive yeast cells (e.g. *Saccharomyces cerevisiae*) and mammalian tumor cells ethanol and lactate excretion occurs, respectively, even if ample oxygen is available.

Investigations into the control mechanisms of glucose degradation by *S. cerevisiae*, using high-performance chemostat-techniques, revealed that ethanol excretion is the consequence of an overflow reaction in the glycolytic pathway from the point of which pyruvate is produced. Käppeli and Sonnleitner concluded that this is due to the limited respirative capacity of this yeast [54]. This was the first time that a mechanistic concept was presented that accounted for both the "Pasteur effect" and "Crabtree" or "glucose effect" [112].

It is of interest that tumor cells display a similar behavior with respect to glucose metabolism as in yeast cells. Transformation of normal cells leads to enhanced glucose uptake, glycolytic flux and lactate excretion. The increased glucose transport and glycolytic flux is due to de novo synthesis of glucose transporters, and higher activities of rate-limiting enzymes in the glycolytic pathway, namely HK, PFK-1 and PK.

Maximum rates of glucose oxidation may be regulated by the limited capacity of TCA-cycle and the mitochondrial electron transfer chain. In relation to this, it is important to notice that the respiratory capacity of tumor cells is not or only slightly reduced (about 10 %) as compared to their ancestral primary cells. This adds support for the model proposed by Sonnleitner and Käppeli [112], which explains ethanol formation in baker's yeast in terms of the restricted respiratory capacities and may also be valid for animal cells. However, too few experimental data are at hand which would allow us to verify or falsify this hypothesis.

In relation to this, it is of interest that although the phylogenetic relationship between yeast and tumor cells seems rather distant, there is now some very strong evidence that many basic cell functions have been highly conserved. For instance, transcriptional control, signal transduction and cell cycle control may be governed by homologous genes and molecular events. Therefore, it seems very likely that the basic principles involved in the metabolic control of sugar degradation were conserved as well. From this we conclude that in animal cells the control of carbohydrate metabolism may share much in common with yeast cells. Attempts to verify this hypothesis will give new insights into basic knowledge, applied research and exploitation of a promising resource. Furthermore, it will be possible to learn more about the process of carcinogenesis.

9 Acknowledgement

We thank Dr. Juliet Lee for revising the manuscript.

10 Abbreviations

ADP	adenosine diphosphate	
AMP	adenosine monophosphate	
cAMP	cyclic adenosine monophosphate	
ATP	adenosine triphosphate	
C_6	glucose	
C_2	ethanol, acetic acid	
CMAS	residual biomass	
D	dilution rate	$[h^{-1}]$, $[d^{-1}]$
D_C	critical dilution rate (for washout); $D_C \equiv \mu_{max}$	$[h^{-1}]$, $[d^{-1}]$
D_R	maximum dilution rate for pure oxidative glucose degradation	$[h^{-1}]$, $[d^{-1}]$
DAG	diacylglycerol	
e	ethanol concentration	$[g\,l^{-1}]$
EGF	epidermal growth factor	
ETOH	ethanol	
F 1,6-P	fructose 1,6-biphosphate	
F 2,6-P	fructose 2,6-biphosphate	
G1	gap 1, cell division cycle	
G2	gap 2, cell division cycle	
Gp	GTP-binding protein	
G 6-P	glucose 6-phosphate	
GLC	glucose	
HK	hexokinase	
IP_3	inositol triphosphate	
M	mitosis, cell division cycle	
μ	specific growth rate	$[h^{-1}]$, $[d^{-1}]$
μ_{max}	maximum specific growth rate	$[h^{-1}]$, $[d^{-1}]$
P_i, PP_i	inorganic phosphate	
PDGF	platelet derived growth factor	
PDH	pyruvate dehydrogenase	
PFK-1	6-phosphofructo 1-kinase	
PFK-2	6-phosphofructo 2-kinase	
PIP_2	phosphoinositol 4,5-biphosphate	
PK	pyruvatekinase	
$PK-M_2$	pyruvatekinase, isoenzyme	

q_{CO_2}	specific carbon dioxide formation rate	[mmol g^{-1} h^{-1}], [d^{-1}]
q_{O_2}	specific oxygen uptake rate	[mmol g^{-1} h^{-1}], [d^{-1}]
$q_{O_{2max}}$	maximum specific oxygen uptake rate	[mmol g^{-1} h^{-1}], [d^{-1}]
q_P (lactate)	specific lactate formation rate	[d^{-1}]
q_P (t-PA)	specific t-PA formation rate	[d^{-1}]
q_S	specific glucose uptake rate	[mmol g^{-1} h^{-1}], [d^{-1}]
$q_{S_{max}}$	maximum specific glucose uptake rate	[mmol g^{-1} h^{-1}], [d^{-1}]
Q	specific flux	
QE	specific ethanol uptake/excretion rate	
QG,	specific glucose uptake rate	
QGG	glycolytic flux	
QGO	glucose flux through TCA-cycle and mitochondrial respiratory chain	
QT	rate of storage material formation/consumption	
RQ	respiratory quotient; q_{O_2}/q_{CO_2}	[—]
S	DNA-synthesis, cell division cycle	
S	residual glucose (substrate) concentration	[mmol l^{-1}, g l^{-1}]
S_0	glucose (substrate) concentration in medium	[mmol l^{-1}, g l^{-1}]
TCA-cycle	tricarbonic acid cycle, citric acid cycle	
TREH	storage materials	
t-PA	tissue-type plasminogen activator	
X	biomass concentration	[g l^{-1}]
$Y_{X/S}$	substrate yield coefficient	[—]

11 References

1. Allard WJ, Lienhard GE (1985) J. Biol. Chem. 260: 8668
2. Arreguín M (1986) Ph D. thesis no. 8089, ETH-Zürich
3. Arreguín M, Kaeppeli O (1987) J. Gen. Microbiol. 133: 2517
4. Barbacid M (1987) Annual Rev. Biochem. 56: 779
5. Baserga R, Surmacz E (1987) Biotechnology. 5: 355
6. Bell RM (1986) Cell. 45: 631
7. Berridge MJ (1986) Inositol lipids and cell proliferation. In: Kahn P, Graf T (eds) Oncogenes and growth control, Springer, Berlin Heidelberg New York, p 145
8. Berridge MJ (1987) Annual Rev. Biochem. 56: 159
9. Berridge MJ (1987) Biochim. Biophys. Acta. 907: 33
10. Birnbaum MJ, Haspel HC, Rosen OM (1987) Science. 235: 1495
11. Bishop JM (1987) Trends in oncogenes. In: Bradshaw RA, Prentis S (eds) Oncogenes and growth control, Elsevier, Amsterdam, p 1
12. Bissell MJ, White RC, Hatie C, Bassham JA (1973) Proc. Natl. Acad. Sci. USA. 70: 2951
13. Bohmann D, Bos TJ, Admon A, Nishimura T, Vogt PK, Tjian R (1987) Science 238: 1386
14. Breeden L, Nasmyth K (1987) Nature. 329: 651
15. Burk D, Schade AL (1956) Science. 124: 270
16. Burk D (1939) Cold Spring Harbour Symposia Quant. Biol. 7: 420
17. Chance B (1959) Quantitative aspects of the control of oxygen utilization. In: Wolstenholme OBE, O'Connor CM (eds) CIBA Foundation Symposium on the regulation of cell metabolism, Churchill, London, p 91
18. Claus R, Käppeli O, Fiechter A (1983) FEMS Microbiology Letters. 18: 185

19. Claus TH, El-Maghrabi MR, Regen DM, Stewart HB, McGrane M, Kountz PD, Nyfeler F, Pilkis J, Pilkis SJ (1984) Curr. Top. Cell. Regul. 23: 57
20. Cooper JA, Esch FS, Taylor SS, Hunter T (1984) J. Biol. Chem. 259: 7835
21. Cooper JA, Reiss NA, Schwartz RJ, Hunter T (1983) Nature. 302: 218
22. Crabtree HG (1929) Biochemical J. 23: 536
23. Darnell J, Lodish H, Baltimore D (1986) Molecular cell biology, Scientific American Books, New York, p 1045
24. DeFeo-Jones D, Scolnick EM, Koller R, Dhar R (1983) Nature. 306: 707
25. DenHollander JA, Ugurbil K, Brown TR, Bedar M, Redfield C, Shulman RG (1986) Biochemistry. 25: 203
26. Den Hollander JA, Ugurbil K, Shulman RG (1986) Biochemistry. 25: 212
27. Denhardt DT, Edwards DR, Parfett CLJ (1986) Biochim. Biophys. Acta. 865: 83
28. Dietzel C, Kurjan J (1987) Cell. 50: 1001
29. Eigenbridt E, Fister P, Reinacher M (1985) New perspectives on carbohydrate metabolism 2, in tumor cells. In: Beitner R (ed) Regulation of carbohydrate metabolism 2, Chemical Rubber Company, Cleveland, p 141
30. Elliott SG, McLaughlin CS (1983) Prog. Nucleic Acid Res. Mol. Biol. 28: 143
31. Fasano O (1986) RAS genes and growth control in the yeast *Saccharomyces cerevisiae*. In: Kahn P, Graf T (eds) Oncogenes and growth control, Springer, Berlin Heidelberg New York, p 200
32. Fiechter A, Fuhrmann FG, Käppeli O (1981) Adv. Microbial Phys. 22: 123
33. Fleischman LF, Chahwala SB, Cantley L (1986) Science. 231: 407
34. Flier JS, Mueckler MM, Usher P, Lodish HF (1987) Science. 235: 1492
35. Gallwitz D, Donath C, Sander C (1983) Nature. 306: 704
36. Gilman AG (1987) Annual. Rev. Biochem. 56: 615
37. Gilman AG (1984) Cell. 36: 577
38. Gosalvez M, Lopez-Alarcon L, Garcia-Suarez S, Montalvo A, Weinhouse S (1975) Eur. J. Biochem. 55: 315
39. Gschwend-Petrik M (1983) Ph.D. thesis no. 7441, ETH-Zürich
40. Hanafusa H (1986) Activation of the c-*scr* gene. In: Kahn P, Graf T (eds) Oncogenes and growth control, Springer, Berlin Heidelberg New York, p 100
41. Hanley MR, Jackson T (1987) Nature. 328: 668
42. Hatanaka M (1974) Biochim. Biophys. Acta. 355: 77
43. Hatanaka M, Hanafusa H (1970) Virology. 41: 647
44. Hatanaka M, Huebner RJ, Gilden RV (1969) J. Natl. Cancer Inst. 43: 1091
45. Hayles J, Nurse P (1986) J. Cell Science. Suppl. 4: 155
46. Hediger MA, Coady MJ, Ikeda TS, Wright EM (1987) Nature. 330: 379
47. Hertz R, Bar-Tana R (1986) Arch. Biochem. Biophys. 250: 54
48. Hunter T (1987) Cell. 50: 823
49. Janshekar H (1979) Ph.D. thesis no. 6402, ETH-Zürich
50. Janshekar H, Fiechter A (1979) Eur. J. Appl. Microbiol. Biotechnol. 6: 341
51. Johnson MJ (1941) Science. 94: 200
52. Jones MN, Nickson JK (1981) Biochim. Biophys. Acta. 650: 1
53. Käppeli O, Arreguín M, Fiechter A (1985) J. Gen. Microbiology. 131: 1411
54. Käppeli O, Sonnleitner B (1986) CRC Crit. Rev. Biotechnol. 4: 299
55. Käppeli O (1986) Adv. Microbial Physiol. 28: 181
56. Kahn P, Graf T (1986) Oncogenes and growth control. Springer, Berlin Heidelberg New York
57. Kaibuchi K, Kikkawa U, Takai Y, Nishizuka Y (1984) Calcium and phospholipid turnover in signal transduction. In: Cohen P (ed) Molecular aspects of cellular regulation, Elsevier, Amsterdam, vol 3 p 81
58. Krebs EG, Beavo JA (1979) Ann. Rev. Biochem. 48: 923
59. Krebser U (1987) Ph.D. thesis no. 8213, ETH-Zürich
60. Krebser U, Adler I, Fiechter A (1987) Chem. Ing. Tech. 59(6): 514
61. Küenzi MT (1970) Ph.D. thesis no. 4565, ETH-Zürich
62. Küenzi MT, Fiechter A (1969) Arch. Mikrobiol. 64: 396
63. Küenzi MT, Fiechter A (1972) Arch. Mikrobiol. 84: 254
64. Lee, MG, Nurse P (1987) Nature. 327: 31

65. Leist C (1988) Ph.D. thesis no. 8537, ETH-Zürich
66. Leist C, Meyer HP, Fiechter A (1986) J. Biotechnol. 4: 235
67. Lentzen H, Arreguín M, Käppeli O, Fiechter A, Fuhrmann GF (1987) J. Biotechnol. 6: 281
68. Lorencez-Gonzales I (1985) Ph.D. thesis no. 7781, ETH-Zürich
69. Lynen F, Hartmann G, Netter KF, Schuegraf A (1959) Phosphate turnover and Pasteur effect. In: Wolstenholme OBE, O'Connor CM (eds) CIBA Foundation Symposium on the regulation of cell metabolism, Churchill, London, p 256
70. Lynen F, Koenigsberger R (1951) Justus Liebigs Annalen der Chemie. 573: 60
71. Lynen F (1941) Justus Liebigs Annalen der Chemie. 516: 120
72. Lynen F (1942) Naturwissenschaften. 30: 398
73. Marx JL (1987) Science. 235: 974
74. Marx JL (1987) Science. 237: 854
75. Mechsner B (1987) Diploma-thesis, ETH-Zürich
76. Mellor J, Kingsman AJ, Kingsman SM (1986) Yeast. 2: 145
77. Meyer C, Beyeler W (1984) Biotechnol. Bioeng. 26: 916
78. Meyer H-P, Leist C, Fiechter A (1984) J. Biotechnol. 5: 355
79. Miyajima I, Nakafuko M, Nakayama N, Brenner C, Miyajima A, Kaibuchi K, Kenichi A, Kaziro Y, Matsumoto K (1987) Cell. 50: 1011
80. Mueckler M, Caruso C, Baldwin SA, Panico M, Blench I, Morris HR, Allard WJ, Lienhard GE, Lodish HF (1985) Science. 229: 941
81. Murray AW (1987) Nature. 327: 14
82. Müller F, Brühl KH, Greidel K, Kowallik KV, Ciriacy M (1987) Mol. Gen. Genet. 207: 421
83. Nishizuka Y (1984) Nature. 308: 693
84. Nurse P (1987) Cell cycle control genes in yeast. In: Bradshaw RA, Prentis S (eds) Oncogenes and growth control, Elsevier, Amsterdam, p 275
85. Parker PJ, Ullrich A (1986) Protein kinase C. In: Kahn P, Graf T (eds) Oncogenes and growth control, Springer, Berlin Heidelberg New York, p 154
86. Pasteur L (1861) Bull. Soc. Chim. Paris. June 28: 79
87. Pedersen P (1978) Prog. Exp. Tumor Rev. 22: 190
88. Petrik M, Käppeli O, Fiechter A (1983) J. Gen. Microbiol. 129: 43
89. Preiss J, Loomis CR, Bishop WR, Stein R, Niedel JE, Bell RM (1986) J. Biol. Chem. 261: 8597
90. Presek P, Glossmann H, Eigenbrodt E, Schoner W, Rübsamen H, Friis RR, Bauer H (1980) Canc. Res. 40: 1733
91. Quafie CJ, Pinkert CA, Ornitz DM, Palmiter RD, Brinster RL (1987) Cell. 48: 1023
92. Racker E (1974) Mol. Cell. Biochem. 5: 17
93. Racker E (1976) A new look at mechanisms in bioenergetics. Academic, New York
94. Randle PJ, Fatania HR, Lau KS (1984) Regulation of the mitochondrial branchedchain 2-oxoacid dehydrogenase complex of animal tissue by reversible phosphorylation. In: Cohen P (ed) Molecular aspects of cellular regulation. Enzyme regulation by reversible phosphorylation-further advances, Elsevier, Amsterdam, vol 3 p 1
95. Rasmussen H (1986) N. Engl. J. Med. 314: 1094
96. Rasmussen H (1986) N. Engl. J. Med. 314: 1164
97. Reibstein D, den Hollander JA, Pilkis SJ, Shulman RG (1986) Biochem. 25: 219
98. Reiling HE, Laurila H, Fiechter A (1985) J. Biotechnol. 2: 191
99. Reuveny S, Velez D, Macmillan JD, Miller L (1987) Develop. Biol. Standard. 66: 169
100. Rieger M (1983) Ph.D. thesis no. 7264, ETH-Zürich
101. Rüther U, Garber C, Komitowski D, Müller R, Wagner EF (1987) Nature. 325: 412
102. Salminen A, Novick PJ (1987) Cell. 47: 413
103. Salter DW, Baldwin SA, Lienhard GE, Weber MJ (1982) Proc. Natl. Acad. Sci. USA. 79: 1540
104. Schatzmann H (1975) Ph.D. thesis no. 5504, ETH-Zürich
105. Schmitt HD (1986) Cell. 47: 401
106. Short NJ (1987) Nature. 330: 209
107. Sibley DR, Benovic JL, Caron MG, Lefkovitz RJ (1987) Cell. 48: 913
108. Singh M, Singh VN, August JT, Horecker BL (1974) Arch. Biochem. Biophys. 165: 240
109. Singh NV, Singh M, August JT, Horecker BL (1974) Proc. Nat. Acad. Sci. USA. 71: 4129
110. Slamon DJ (1987) N. Engl. J. Med. 317: 955

111. Sols A, Gancedo C, Dela Fuente G (1971) Energy-yielding metabolism in yeasts. In: Rose AH, Harrison JS (eds) The yeasts, Academic, New York, vol 2 p 271
112. Sonnleitner B, Käppeli O (1986) Biotechnol. Bioeng. 28: 927
113. Spector M, O'Neal S, Racker E (1981) J. Biol. Chem. 256: 4219
114. Spector M, Pepinsky RB, Vogt VM, Racker E (1981) Cell. 25: 9
115. Struhl K (1987) Cell. 50: 841
116. Strässle C (1988) Modell zur Spontansynchronisation von Saccharomyces cerevisiae. Ph.D. thesis no. 8598, ETH-Zürich
117. Strässle C, Sonnleitner B, Fiechter A (1988) J. Biotechnol. 7: 299
118. Sugano S, Hanafusa H (1985) Mo. Cell. Biol. 5: 2399
119. Tanner W, Gallwitz D (1986) Cell cycle and oncogenes. Springer, Berlin Heidelberg New York
120. Uhlinger DJ, Yang C-Y, Reed LJ (1986) Biochemistry. 25: 5673
121. Van Schaftingen E (1987) Advances in Enzymology. 59: 315
122. Varmus HE (1987) Science. 238: 1337
123. Verma IM, Saasone-Corsi P (1987) Cell. 51: 513
124. Vitkauskas GV, Canellakis ES (1974) Biochim. Biophys. Acta. 823: 19
125. Wagner B (1987) Ph.D. thesis no. 8225, ETH-Zürich
126. Wakelam MJO, Davies SA, Housley MD, McKay I, Marshall CJ, Hall A (1986) Nature. 323: 173
127. Warburg O (1924) Biochem. Z. 152: 309
128. Warburg O (1925) Sonderdruck aus klin. Wochenschrift München. 4: 1
129. Warburg O (1926) Über den Stoffwechsel der Tumoren. Springer-Verlag, Berlin
130. Warburg O (1956) Science. 123: 309
131. Warburg O (1956) Science. 124: 269
132. Weinberg RA (1987) Cellular oncogenes. In: Bradshaw RA, Prentis S (eds) Oncogenes and growth control, Elsevier, Amsterdam, p 11
133. Weinhouse S (1956) Science. 124: 267
134. Wheeler TJ, Hinkle PC (1985) Annual Rev. Physiol. 47: 503
135. Wilson JE (1985) Regulation of mammalian hexokinase activity. In: Beitner R (ed) Regulation of carbohydrate metabolism, Chemical Rubber Company, Cleveland, vol 1 p 45
136. Witters LA, Vater CA, Lienhard GE (1985) Nature. 315: 777
137. Wyke JA, Stoker AW (1987) Biochim. Biophys. Acta. 907: 47
138. Yoshimasa T, Sibley DR, Bouvier M, Lefkowitz RJ, Caron MC (1987) Nature. 327: 67

Bioreactor for Mammalian Cell Culture

Ales Prokop[1] and Morris Z. Rosenberg[2]
[1] Washington University, St. Louis, U.S.A.; [2] Invitron Corporation, St. Louis, USA

Purpose of this article is to review the current status of bioreactor design for mammalian cell culture. Morphological and biochemical features of two major mammalian cell groups, anchorage-dependent and independent cells are proposed as a basis for different behavior at their cultivation. Different bioreactor configurations are systematically discussed through enumerating elementary physical phenomena and through stressing their physiological significance. Special considerations are given to those areas which are inherent to mammalian cell bioreactor.

Advances in Biochemical Engineering/
Biotechnology, Vol. 39
Managing Editor: A. Fiechter
© Springer-Verlag Berlin Heidelberg 1989

1 Introduction

The bioreactor design for mammalian cell cultures is an area undergoing rapid development. Mammalian cultures have been tradionally used for production of vaccines. The exploitation of recombinant DNA technology has entered a practical stage, giving rise to a number of high value protein products with significant commercial potential.

The practical employment of mammalian cultures is hampered by the fact that these cells require rather narrow chemical and physical environmental conditions [1]. The strict control of such conditions is possible through detailed analysis of cells' requirements. No attempt has been made to comprehensively cover all aspects of mammalian bioreactor design, although some reviews of rather qualitative nature exist (e.g. Refs. [2-4]). It is the intention of this paper to discuss important aspects of bioreactor design with proper balance between qualitative and quantitative treatment, i.e. elementary physical phenomena will be presented in the context of their physiological conditions. Some other phenomena are presented only in terms of differences between traditional bioreactor and those used in the mammalian cell culture area. The article concludes with an outlook for development in the areas requiring further attention.

2 Mammalian Cell Morphology and Technological Implications

2.1 Lipid Bilayer and Cytoskeleton

Mammalian cells are surrounded by a phospholipid bilayer membrane, embedded with enzymes and structural proteins, which mediates communication between the cell and the environment. These cells lack an outer cell wall and, as a result, are highly sensitive to environmental stimuli, such as osmotic changes, hydrodynamic forces, pH and nutrient changes. The lipid bilayer membranes are essentially and elastic (deformable) structures. Some specialized proteins in membrane are involved in accepting different external stimuli, such as chemical, osmotic, mechanical, etc., e.g. receptors, ion channel proteins [5, 6]. Most membrane proteins are attached to microfilament proteins of the cytoskeleton (see below), such as spectrin. This attachment limits receptor lateral mobility in the membrane [7] and for certain class of receptors their internalization upon presentation of an external stimuli (chemical ligand, mechanical). The receptor/ligand internalization is accomplished through an endocytosis, followed for some classes by recycling through an exocytosis. Besides that receptor/ligand complex can also undergo a turnover (degradation) [8]. A damage to receptors through mechanical shear (or simple dislocation) may have a vital role for a cell as the signals from the environment will not be properly propagated into the cells interior. Schwartz et al. [9] noted an increased low-density lipoprotein (LDL) receptors internalization and degradation at a moderate shear stress of 30 dynes cm^{-2}.

In mammalian cells there are three types of cytoplasmic fibers making up the cell cytoskeleton and are embedded in a dense, highly viscous cytoplasmic gel [10, 11]: the microtubules, which are 25 nm in diameter and possibly hollow, the intermediate filaments (10 nm in diameter) and the microfilaments (6 nm in diameter). The total number of cytoskeleton-associated proteins is on the order of 1,000.

A rigid framework of microtubules, composed of tubulin, is responsible for maintaining the reciprocal position of cytoplasmic components, particularly of lysosomes and mitochondria. Microtubular network is usually close to a nucleus and anchored to a material associated with the centriole. A contractile system is made up of microfilaments, composed of spectrin, actin and of other proteins. Filaments, usually in form of thick bundles (stress fibres) are localized on the inner side of the plasma membrane across the cell and connect the membrane to the nucleus. The attachment to the plasma membrane phospholipid bilayer is via specialized proteins. A pseudo-podial extension of plasma membrane (microvilli), mediated through stress fibers [12], provides better surface area to volume ratio for nutrient uptake, reception, attachment and movement. Both spectrin (actin) and tubulin are formed by polymerization of protein subunits and their dissociation and reassembly are closely regulated in the cell [13]. Intermediate filaments are linked to cell differentiation and may hold the cell nucleus in position. They are the most stable components of the cytoskeleton and the least soluble. However, the intermediate filament network of *Drosophila* was shown to be very sensitive to number of stresses (e.g. heat), and collapses around the nucleus shortly after a heat shock. The normal morphology is resumed when cells are allowed to recover from the stress [14].

Cells in primary or subcultured (non-transformed) cell cultures are of both adherent (anchorage-dependent) and non-adherent (anchorage-independent) types. The latter type grows in a suspension culture. This ability is restricted to hemopoietic cells and in a limited degree to cells derived from many normal tissues, the identity of whose remain unclear. Perhaps, they result from the stem cell or from uncomitted precursor cell compartment [15]. Most of the non-transformed cell lines are, however, of adherent type.

The initial contact between cells or between cells and a substrate (a surface for adhesion) is via, among others, a morphological pattern known as a tight junction, which may develop into gap junction [16]. Junctions provide a cell-to-cell communication and exchange of small molecules between cells. Similar cell junctions, called adhesion plaques (patches) enable cell adhesion to a substrate. The attachment is mediated via anchor glycoprotein (e.g. fibronectin in fibroblasts) capping, formed as a result of cell surface cross-linking by multivalent substrate [16]. With the help of cytoskeletal elements, attached to glycoproteins, the membrane fluidity is decreased, the lateral movement of anchor glycoproteins is lowered and cell rounding up is prevented. A final outcome of cell adhesion is the cell spreading (flattening).

Many non-malignant cells, particularly fibroblasts, exhibit a typical saturation pattern when grown attached to a substrate. The density-dependent inhibition of growth (topoinhibition) results in monolayer formation, with limited overlapping and multilayering in dense cultures. A detached cell, usually due to mitosis, will loose its ability to grow and proliferate unless reattached [17]. Detached cells are round and arrested in a certain stage of the cell cycle (G1).

Transformed and genetically engineered cells are in many ways different from those of normal cells [15]: they exhibit rounded shape, have increased life-span, increased lateral mobility of surface proteins and alterations in surface proteins (and receptors) and in cytoskeletal elements. Because of the last three changes such cells usually loose a capability to adhere (anchorage-independent cells). Other transformed cell lines are in an intermediate state, i.e. they are able to grow both attached or in a suspension. When attached they grow in multiple layers beyond confluent monolayer stage. This is particularly apparent for those lines capable of moderate attachment to supports like microcarriers, as observed for some genetically-engineered fibroblast cultures. As the transformation is a multistep process [e.g. Ref. [18]] it is conceivable that the type of growth (ranging from normal and mostly attaching type of completely non-attaching type with several intermediate states) would result from a degree of the cell transformation (e.g. of the incorporation of a foreign viral genome). On the other hand, cell types which would normally not attach because of their genetic make-up (lymphocytes and derived hybridomas) can be in some cases adapted for an attachment [19]. The cells' adaptation is explained on the basis of availability of adhesion protein factors [20]. Both transformed and engineered cell lines typically exhibit enhanced growth and proliferation characteristics as compared to normal cell lines, a property of great advantage for biotechnological purposes.

Hybridomas result from cell hybridization between B or T lymphocytes and a suitable indefinite life span cell like lymphomas, myelomas, etc. (transformed cells). The product of fusion is a cell with nuclei from both parent cells (heterokaryon), immediately undergoing mitosis, yielding a mononucleated hybrid cell. In hybrid cells many genes of either parent continue to be expressed [21].

Lymphocytes (and transformed cells) have typically a depolymerized microfilamental cytoskeleton. They are also suspected to have the other two major components of cytoskeleton in less polymerized state [22, 23]. The same would apply to hybrid cells or hybridomas [24]. The hybrids are expected to have morphology of an intermediate type between those of the parents. A limited amount of information is available regarding the cytoskeleton organization in hybridoma cells [25]. Some information on cytoskeleton of lymphocytes resulted from a work on responses to mechanical stimuli [26, 27]. Little is known about the interaction between the three major filament systems of the cytoskeleton as a result of any imposed chemical or mechanical stimuli.

In summary, mammalian cells' bilayer membrane lacks rigidity and thus does not provide satisfactory protection against the outside disturbances. Surface proteins and receptors, embedded in the membrane structure, provide a communication means with the external environment by accepting all kinds of stimuli. Because of an intimate association of surface receptors with the cell cytoskeleton, a response to external stimuli results in cytoskeleton modification or receptor itself is involved in the stimuli transduction. The cells' recovery from any stress is accompanied by reforming of normal morphology. Mechanical stimuli due to hydrodynamic shear (Sect. 4.4) may cause a deformation of cells cytoskeleton, followed by short or long term responses (Sect. 2.2). In hybridoma, the cytoskeleton network is less distinguished as compared to a typical mammalian cell, provides a limited support to keep the cells' shape and contributes to the anchorage-independency of such cells.

2.2 Stimulus-Response Cascade

Several schemes of a stimulus-response cascade in regard to different stimuli have been suggested in the past [11, 28]. We are of the opinion that all schemes can be generalized into a phenomenological model, applicable to all kinds of cells and stimuli, including chemical and mechanical ones. The postulated sequence of events is as follows [29]:

1) Signal reception takes place through specialized surface membrane proteins, receptors, and cytoskeleton [6, 30];
2) signal transduction and amplification occurs either via second messengers (calcium ions, diacylglycerol, inositol phospholipids, etc. [31]) or directly via an activation of membrane protein phosphorylation through protein kinase C [32];
3) short-term transient response manifests in terms of gene activation (or inhibition), protein expression (e.g. of the specific protein group due to a given stimuli like heat shock proteins, Ref. [33]) and stimulation of metabolism [34]; and
4) in the long-term, numerous cellular responses such as differentiation, etc., show-up [35].

The knowledge of the mechanism of response of mammalian cells to different kinds of external stimuli is important for establishing optimal production conditions. External stimuli (stresses) may include temperature shock, glucose starvation, lack of oxygen, exposure the heavy metals, mechanical stress, etc. The short-term responses (expression of some specific stress protein with simultaneous shutdown of overall protein biosynthesis) may serve to help a cell to recover from the unfavorable situation and may thus provide a necessary homeostatic function [35]. The long-term responses have a relevance to synthesis of targeted product as the cells physiological status (differentiation) is important in that respect.

Table 1 lists some possibilities of stress induced protein synthesis in mammalian cells. The synthesis of specific stress proteins occurs at the expense of other proteins. The length of stress exposure to induce a response is usually between 0.5–3.0 h. Parag et al. [38] reported a peak in response in 0.5 to 2 h after the exposure for temperature-induced transient protein synthesis, although it may take as long as 7 to 8 h [14]. The recovery may take up to 12 h. Obviously, the response, peaking and recovery are functions of the cell cycle length. Thus, in bacteria a response is noted in 5 to 10 min following a alkaline shift [37]. Generally, the transient disturbances in normal protein synthesis will not have a serious technological consequence, unless short-term fluctuations (e.g. due to a controller on-off periods) are integrated. It is not, however, clear how cells perceive a variety of frequencies and amplitudes of a given disturbance. This will obviously require further research.

Table 1. Some possibilities of stress induced protein synthesis in mammalian cells

Stimuli	Class of protein(s)	Ref.
Heat	Heat shock	33)
Anoxia	Anaerobic stress	35)
Glucose starvation	Glucose-regulated	36)
Other nutrient starvation	To be expected	—
Mechanical stress	Hydrodynamic shear stress proteins (our preliminary data)	29)
pH	Alkaline shift proteins (possibly by analogy with other organisms)	37)

3 Mammalian Cell Culture Methods

3.1 Type of Support for Anchorage-Dependent Cells

In the following cell culture methods will be discussed from two different viewpoints. First, type of support will be considered (for anchorage-dependent cells) and second, the proposed reactor classification will provide a basis for discussion of culturing of both anchórage dependent and independent cells.

Spier and Fowler [39] have divided cultivation methods for anchorage dependent cells according to two different approaches. The one, where the scale-up is achieved via an increase in the number (or multiplicity) of units and is defined as a *multiple* process; the second, where scale-up is accomplished via building a larger unit, called a *unit* process. Advantages of multiple processes are: a) a contamination of one unit does

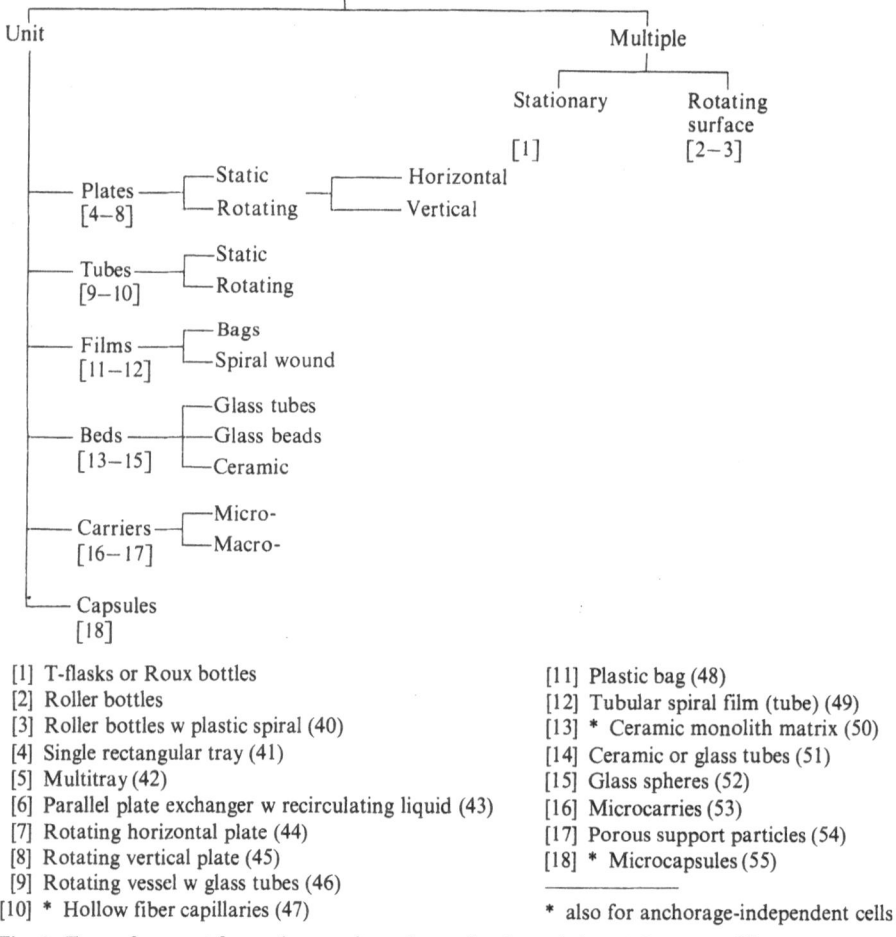

[1] T-flasks or Roux bottles
[2] Roller bottles
[3] Roller bottles w plastic spiral (40)
[4] Single rectangular tray (41)
[5] Multitray (42)
[6] Parallel plate exchanger w recirculating liquid (43)
[7] Rotating horizontal plate (44)
[8] Rotating vertical plate (45)
[9] Rotating vessel w glass tubes (46)
[10] * Hollow fiber capillaries (47)

[11] Plastic bag (48)
[12] Tubular spiral film (tube) (49)
[13] * Ceramic monolith matrix (50)
[14] Ceramic or glass tubes (51)
[15] Glass spheres (52)
[16] Microcarries (53)
[17] Porous support particles (54)
[18] * Microcapsules (55)

* also for anchorage-independent cells

Fig. 1. Type of support for anchorage-dependent cell culture (adapted from Ref. [39])

Table 2. Classification of mammalian cell bioreactor according to reactor type

Reactor type	Type of support/substrate	Suspension (S) Anchored (A)	Ref.
1. Stirred tank reactor	Possible for all:	A and S	[57]
1.1. Flat turbine	Microcarriers	A and S	[57]
1.2. Stirrer bar/paddle	Microcapsules	A and S	[41]
1.3. Paddle w flat sheet	Porous support	A and S	[41]
1.4. Angled blades	Beads	A and S	[57]
1.5. Marine propeller		A and S	[57]
1.6. Vibromixer	Except: not recommended for microcarriers	A and S	[58]
1.7. Plough-shaped sheet		A and S	[57]
1.8. Helical mixer		A and S	[57]
1.9. Anchor mixer		A and S	[57]
1.10. Helical w draft tube		A and S	[57]
1.11. Ribbon and screw		A and S	[57]
1.12. Cavity (cage) type		A	[59]
1.13. Hanging stirrer		A and S	[60]
1.14. Float type		A and S	[61]
1.15. Horizontal loop type		A and S	[62]
1.16. Film type		A and S	[63]
2. Column reactor			
2.1. Column	Rotating vertical stack plate	A	[45]
	Rotating horizontal disc	A	[44]
2.2. Bubble column	Porous support or beads	A and S	[64]
2.3. Column w draft tube	Porous support or beads	A and S	[64]
2.4. Fluidized bed (simple column or staged)	Porous support or beads	A	[54]
2.5. Packed bed			
2.5.1.	Glass spheres	A	[52]
2.5.2.	Ceramic or glass tubes	A	[51]
3. Membrane reactor			
3.1 Hollow fiber	Capillary (extracapillary or lumen)	A and S	[47]
3.2. Flat membrane			
3.2.1. Dialysis reactor	No support	S	[65]
3.2.2. Dialysis bag/fleaker	No support	S	[66]
3.2.3. Maintenance reactor	No support	S	[67]
(membranes for	Porous support	A	
nutrients supply)	Microcarriers	A	
3.2.4. Ceramic monolith	Ceramic support	A and S	[50]
3.3. Microencapsulation	Capsule	A and S	[55]
3.4. Gel matrix	Gel bead	A and S	[68]

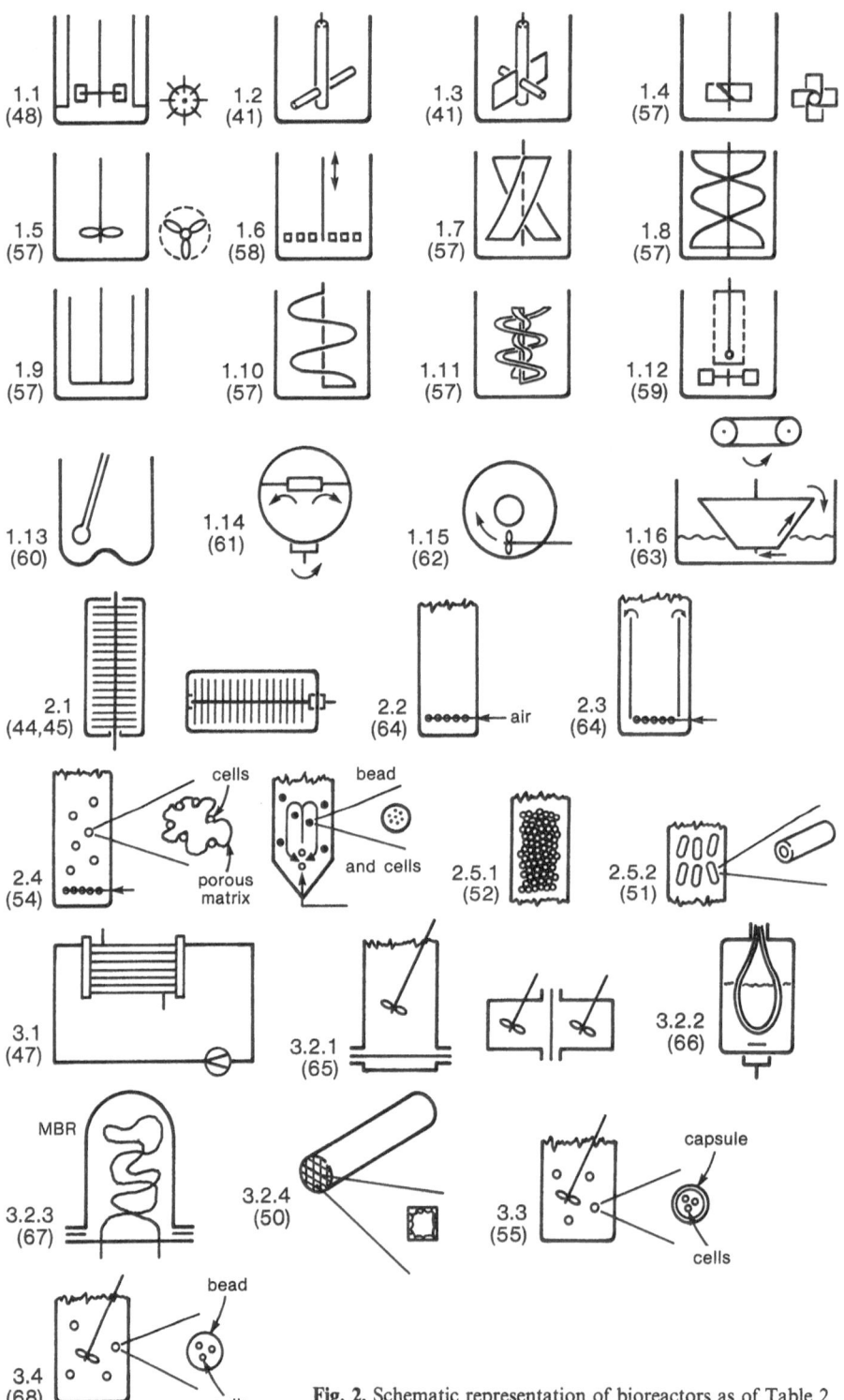

Fig. 2. Schematic representation of bioreactors as of Table 2

not ruin others; b) scale-up is achieved via the employment of a number of identical units; and c) equipment is simple in design. Advantages of unit processes are: a) they are labor and space intensive; b) better process monitoring, control and quality is achieved; c) contamination control is easier, as there are less handling operations; and d) it provides a moderate scale-up potential in some cases [39]. Figure 1 depicts types of support for anchorage-dependent cells.

Multiple processes have been extensively reviewed by Panina [56], allowing for mono- and multi-layer growth type. Rotating roller bottle provides better liquid renewal and aeration, and consequently higher cell yields. Depending on the cell and support type, monolayer, multilayer or/and bridging of cells are encountered. Unit processes were summarized by Spier and Fowler [39]. From the industrial viewpoint, the most important unit processes are those involving microcarriers and micro-capsules. Among emerging cultivation methods ceramic matrix and hollow fiber capillaries are important. Roller bottles are still in use in some pharmaceutical companies as an example of multiple processing.

3.2 Reactor Types

Another classification, based on reactor type, encompasses both anchorage-dependent and -independent cells. Table 2 presents more practical devices of the unit type processing, most of them of industrial importance. Their schematic representation is in Fig. 2 with the numbering adopted from Table 2. Figure 3 presents possible reactor bottom design applicable for stirred reactors.

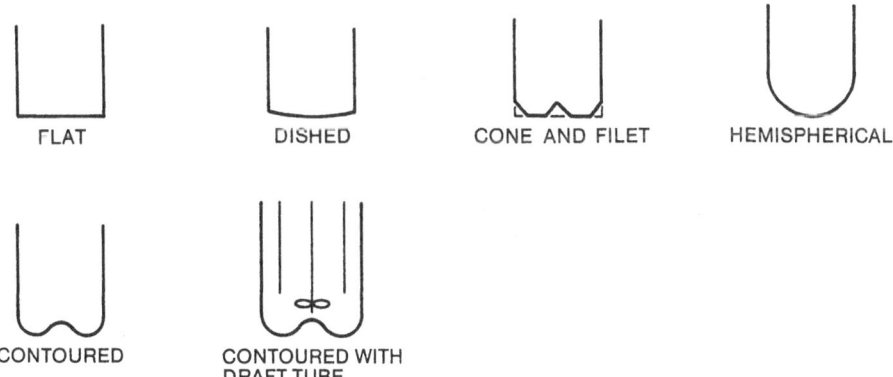

Fig. 3. Types of bottom design for stirred tank reactors (Refs. [60,57,92])

4 Bioreactor Configurations and Their Characterization

4.1 Flow Pattern and Macromixing

4.1.1 Mechanically-Agitated Reactors

The axial type of flow is the preferred pattern for mammalian cells as contrasted with radial type of flow. The predominant radial flow with rather local energy distribution is still used in some cases (type 1.1–1.3 of Table 2). All kinds of axial mixers have been proposed (types 1.4–1.14 of Table 2). Vibromixer has been introduced by Chemap [58]. Helical mixer with a small pitch (sail type) has been applied by Monsanto [67,69]. The helical mixer [70] (alternatively with a draft tube [71]), anchor, ribbon and screw types are prospective and are to be tested in particular applications. They all provide rather low shear mixing and energy is more uniformly distributed. New Brunswick Scientific, Inc. has introduced a cavity type of axial mixer with aerated space separated from the cell compartment [59,72]. Cells are attached to microcarriers and excluded from the aerated space. Hanging and float type mixers were introduced by Techne, Inc. [60,61]. Flow patterns of most of the above configurations are discussed in Oldshue [57]. Bajpai and Reuss [73] presented a comprehensive review on characterization of mixing in mechanically agitated bioreactors with an attempt to identify vessel compartments and their interplay. Last two types (1.15–1.16) could be utilized for mammalian cells because of their low shear mixing fields. Horizontal loop type has been introduced for mixing of viscous media by Läderach et al. [62] and film reactor by Roubicek and Feres [63]. The latter type is suitable for both anchored and suspended cells.

4.1.2 Pneumatically-Agitated (Column) Reactors

Flow patterns have been well established for different column-type reactors, particularly of type 2.2. and 2.3. (Table 2). The type 2.3. is denoted usually as an air-lift reactor with two basic alternatives: draft tube aeration and annular aeration. A review of Erickson and Stephanopoulos [64] and a paper of König et al. [74] may represent numerous articles characterizing mixing pattern in bubble and air-lift columns. Fluidized-bed (type 2.4.) with a tapered or conical bottom [75] is particularly useful for cells immobilized or entrapped in beads and capsules. Bellco Glass, Inc. has recently introduced a glass reactor with orderly mixing pattern of alginate beads (with immobilized cells [75]), Verax Corp. uses a highly porous collagen support in a fluidized bed reactor [54]. The supports provide a physical protection to cells entrapped inside.

4.1.3 Membrane Reactors

Hollow fiber bioreactor (Table 2, type 3.1.) is usually connected with the main medium reservoir in a loop-type arrangement and thus the macroflow through the cartridge can be properly controlled. Depending on the construction, two types of convective flow through the culture bed (either in hollow fiber lumen or in the extra-capillary space) can occur: axial flow with longitudinal gradient along the hollow fiber tubes [47,76] or radial (cross-flow) [67]. A flat membrane bioreactor (type 3.2.) exists in several alternatives. A dialysis reactor may have a membrane attached to the

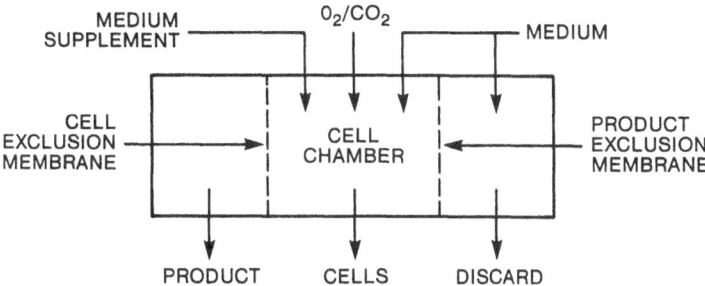

Fig. 4. Schematics of Mass Culture Technology (MCT) membrane bioreactor (adapted from Ref. [65])

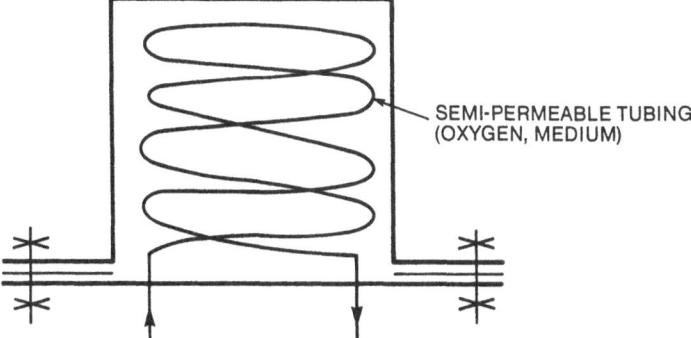

Fig. 5. Schematics of Maintenance Bioreactor (MBR). Semipermeable tubing serves to supply all kinds of nutrients (adapted from Ref. [67])

reactor bottom [77] or between two or three stirred vessels. The third type has been introduced by Bio-Response, Inc. [65] and Klement et al. [78]. Schematics of Bio-Response MCT bioreactor is in Fig. 4. No information is available about the culture compartment mixing. A similar commercial system has been put on the market recently [79]. A dialysis bag or fleaker system [66] provide essentially non-mixed culture compartment. A prospective version of the membrane bioreactor is that of Maintenance Bioreactor (MBR) of Monsanto [67, 80], essentially closely packed bed of cells without mixing provided with semi-permeable tubings for nutrients and oxygen input and product removal (Fig. 5). Cells can be attached to microcarriers or to a porous support. The ceramic monolith support represents essentially a macroporous support for cells. It is a column-type reactor with numerous channels arranged in a recirculation loop (Fig. 6) [50]. Cells (both anchorage dependent and independent) are retained through a physical entrappment and/or due to a charge. Because of a high porosity support, the retention of high-molecular weight products is not achieved in this type of reactor. The microencapsulation (type 3.3., [55]) and gel matrix type (type 3.4., [68]) membrane reactor arrangements are usually materialized in a stirred vessel or in a conical air-lift reactor with well defined macromixing regime. A great

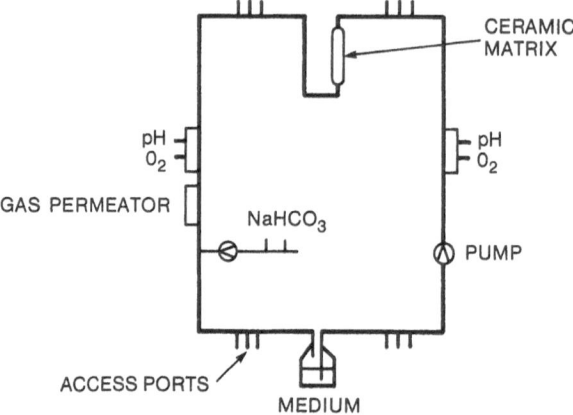

Fig. 6. Schematics of Opticell ceramic monolith macroporous support bioreactor (redrawn from Ref. [49])

number of polymers have been used for this purpose [81]. On the other hand, the gel matrix type of culture does not provide for the product retention.

The only experimental attempt to measure flow pattern of extracapillary immobilized mammalian cells in a hollow fiber culture is that of Tyo and Hu [82], who used a tracer technique.

The advantages of a semi-permeable membrane bioreactor are numeous [3]:

a) High molecular weight products such as monoclonal antibodies and lymphokines (15–450 kDa in size) can be retained within the membrane bound space and separated from other low molecular weight toxic products, such as lactate and ammonium.

b) Toxic metabolic products are continuously removed from the culture space by diffusion through the membrane.

c) Cells are protected from high shear rate/stress in mechanically mixed bioreactor.

d) A high cell density obtained in most membrane bioreactors provides for increased cell-cell contact and interaction, and, presumably for favorable physiological microenvironment suitable for product formation [83].

4.1.4 Operation Modes

From the macroscopic flow viewpoint the bioreactor operation mode can be characterized as follows [84]: a) continuous, usually in a perfusion mode (cells are typically retained), b) fed-batch (also in a perfusion mode), and c) batch (a rather rare mode).

The perfusion mode is usually achieved by a build-in rotating filter allowing for cells retention in the bioreactor [67] or by a filter in an external loop [85]. The latter device is marketed by Sulzer AG, Winterthur, Switzerland. The achieved cell densities are usually several times higher (10^7 cells per ml) compared to a batch mode (2 to 3×10^6 cells per ml). The MBR [67] can be considered as an extreme case of the perfusion mode with cell densities reaching 10^8 cells per ml.

4.2 Suspension Criteria

The following part is of significance in providing a certain microenvironment for high density cultures, and cultures with microcarriers, capsules and beads. This is because of the fact, that the uniformity in suspension is more important than mass transfer.

4.2.1 Unsuspended Solids Criteria

The most frequent criterion is that of the complete off-bottom suspension (CBS)[86]. According to this criterion the suspension speed N_{cs} is defined as a speed to ensure that no solids remain stationary on the bottom of the reactor for no longer than 1 to 2 s. Chudacek[87] proposed a 98% suspension criterion: it is the speed at which 98% of the total solids are suspended. This criterion was suggested to overcome the particular difficulty of disproportionate increase of energy expenditure when suspending the last amount of unsuspended solids. According to Chudacek[88] the existence of induced recirculation loops near the wall-to-bottom junction of the flat-bottom reactor makes the corner area of the reactor difficult to mix (Fig. 7). The unsuspended particulates can be, however, removed by eliminating induced circulation loops[89]

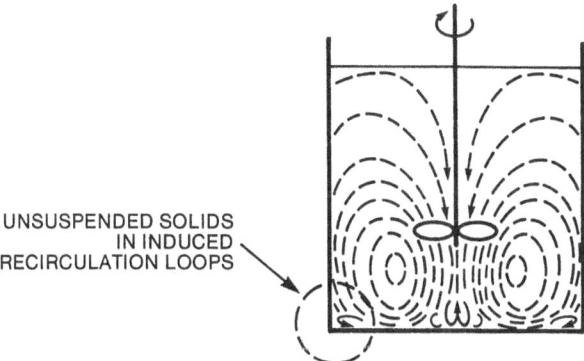

UNSUSPENDED SOLIDS IN INDUCED RECIRCULATION LOOPS

Fig. 7. Schematic flow pattern in a baffled flat bottom tank agitated by propeller featuring recirculation loops at reactor corners (baffles not shown)

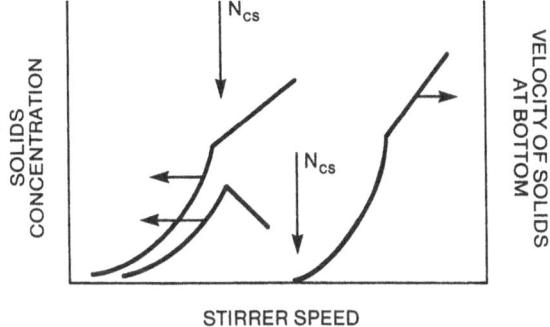

SOLIDS CONCENTRATION

VELOCITY OF SOLIDS AT BOTTOM

N_{cs}

N_{cs}

STIRRER SPEED

Fig. 8. Principle of determination of the stirrer speed for complete suspension (N_{cs}) (Refs.[90-92])

through the use of dished-bottom or profiled-bottom reactors (Fig. 3). The value of CBS criterion is determined usually either visually or via the Doppler velocity measurements of particles at the bottom. A sharp transition in stirred speed to achieve a rise in velocity and consequently the complete particle suspension has been noted (Fig. 8)[90]. Nienow[93] reported on two occasions of a discontinuity in particle concentration with the impeller speed (Fig. 8).

The upward flow velocity associated with the lifting of unsuspended solids will exert two possible forces on particles: drag (lift) and shear near the wall. The contribution of the second force has some practical implications in scale-up. For two identical particulate suspensions suspended in reactors of different sizes, the particles in the smaller reactor would suspend at a lower mean upward velocity than in the larger reactor due to a higher shear rate contribution.

To move a particle from a bottom and lift it, the size of turbulence eddy must be several times larger than the particle size. For particles between 100 μm and 2 mm the size of eddies which lift the particles is in the inertial sub-range of the turbulence spectrum. Based on equivalence of forces at particle lift (neglecting shear) and for conditions of isotropic turbulence Buurman et al.[90] derived a condition for the complete suspension in terms of modified Froude number:

$$\frac{\varrho N_{cs}^2 \, d^{4/3}}{g \, \Delta\varrho \, d_p^{1/3}} = \text{constant} \tag{1}$$

which implies that

$$N_{cs} \propto d^{-2/3} \tag{2}$$

The isotropic turbulence assumes an identity of certain velocity components in three dimensions of a turbulent field[94]. Equation (2) holds only for constant ratio of blade thickness b and impeller diameter d (b/d). Axial stirrer speeds data for complete suspension at two different reactor scales are in agreement with Eq. (2), (Fig. 9).

Inclusion of a draft tube in conjunction with an axial flow impeller greatly enhances particle suspension[94]. Similarly, some improvement in the organized

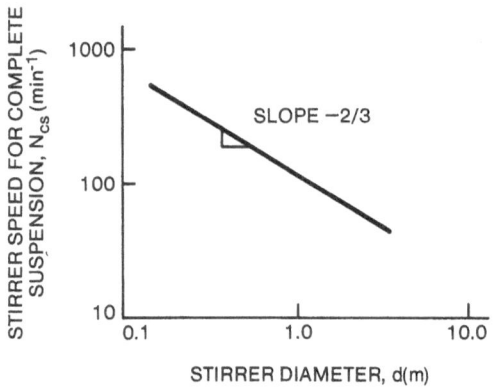

Fig. 9. Stirrer speeds for complete suspension at two scales. Sand in water, $d_p = 157 \, \mu m$, $\varphi = 4.5$ to 14% (adapted from Ref.[90])

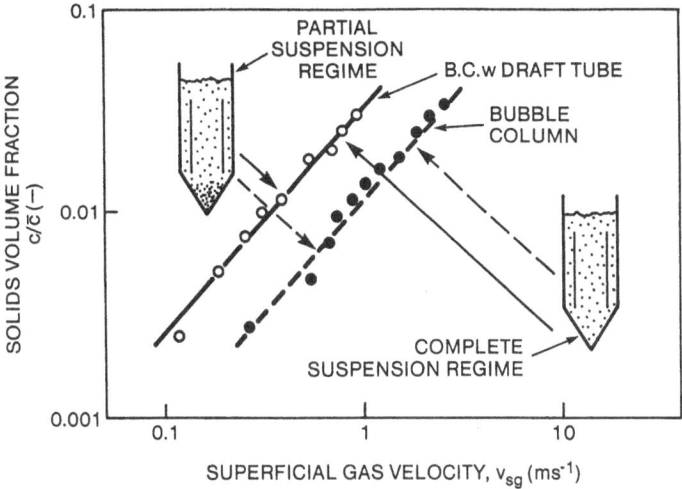

Fig. 10. Flow regimes for solid suspension in bubble column w draft tube as dependent on superficial velocity. Carbon beads in water, $d_p = 460 \, \mu m$, $\varrho_p = 1.3 \, g \, cm^{-3}$, column $H = 900 \, mm$ (adapted from Ref. [95])

structures of flow, such as in a loop reactor or in reactors with a high aspect (H/D) ratio, will lead to a decrease of the exponent in Eq. (2) and bring it close to (-1) [97]. As a consequence a lower N_{cs} would be required to achieve the same degree of suspension. Further improvement is achieved with non-standard vessel geometries. Aeschbach and Bourne [96] reported on minimal energy requirement for a particle suspension for contoured bottom with minimal dead zones (Fig. 3).

Gassing of mixed particulate suspension may, however, complicate the hydrodynamic conditions and invalidate the above observations. In any case, an increased power (higher agitator speed) is required to achieve the same degree of suspension with a gassing as compared to the non-gassed conditions [93].

Aerated column or a bubble column with a draft tube provides another alternative for suspending of solids. Muroyama et al. [95] used a conical bottom column with draft tube to suspend activated carbon beads. The critical gas flow rate was established through a visual observation of solids to lift them from the bottom to just achieve solids suspension. Two regimes were identified, the one with partial and the other with the complete suspension regime (Fig. 10). For both regimes solids concentration in the draft tube was significantly higher than that in the annulus.

4.2.2 Homogeneous Suspension Criteria

To obtain a certain degree of homogeneity in a reactor, not only must the particles be lifted from the bottom of the vessel (as is the case of particle suspension phenomena in Sect. 4.2.1), but they must also be transported through the whole volume of the reactor. Generally, the minimum speed required to achieve homogeneous suspension (N_{hs}) is much greater than that to achieve complete suspension (N_{cs}). It is not the eddies of the inertial subrange but also the largest eddies (i.e. the circulation) that are responsible for the reactor homogeneity. Particularly important are convective

flows near the reactor base. Based on his experimental studies Nienow [93] concluded that the average fluid velocity at any level in an agitated reactor is given by:

$$\bar{n} = k(Nd^2)/(D^2H)^{1/3} \qquad (3)$$

where k is dimensionless coefficient which decreases rapidly with increasing distance
from the impeller. The first conclusion from this equation is that the impeller should
be placed close to the vessel base in order to achieve a velocity at the vessel bottom
(base velocity $\bar{u}_{base}$) with the lowest agitator speed. Secondly, the following relationship
can be derived from Eq. (3) assuming turbulent conditions:

$$\bar{u}_{base} \sim \varepsilon^{1/3} d^{1/3} \qquad (4)$$

which means that $\bar{u}_{base}$ is achieved at a lower value of energy dissipation rate (ε)
by using a large size of impeller d. This is in agreement with experiments for radial
flow agitators. Further advance has been possible with a dimensional analysis.
Thus Buurman et al. [90] derived a simple Froude relationship to define homogeneity,
based on the assumption of fluctuating velocity being proportional to the circulation
velocity (i.e. to the stirrer tip speed):

$$\frac{\varrho N_{hs}^2 d^2}{g \, \Delta\varrho \, d_p} = \text{constant}, \qquad (5)$$

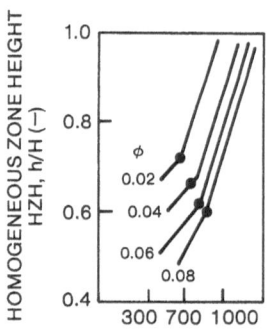

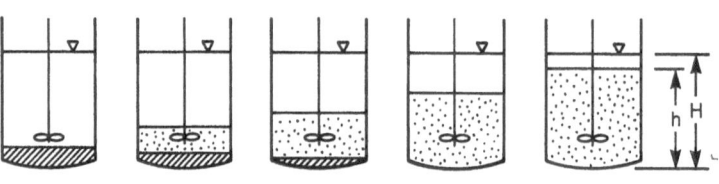

INCREASING SPEED OF ROTATION ⟶

Fig. 11. Variation of the suspended height as a function of stirrer speed. Glass beads in water,
$d_p = 190 \, \mu m$, $\varphi = 6\%$ (v/v) (adapted from Ref. [98])

which implies that the impeller speed to achieve homogeneous suspension is indirectly proportional to impeller diameter:

$$N_{hs} \propto 1/d \quad \text{or} \quad N_{hs}d \propto \text{constant} \tag{6}$$

Equation (6) is a form of a tip speed correlation with N_{hs} substituted for N.

Suspension homogeneity can be estimated through measurements of concentration at different locations in a reactor. Particle segregation during a sample withdrawal often imposes difficulties in obtaining a representative sample. Thus measurement of local variation is conveniently substituted by the height of homogeneous zone (suspended height/interface height) — HZH [99]. Figure 11 illustrates the principle of the method. Note that a breakpoint on curves in Fig. 11 corresponds to the CBS criterion equal to one second for complete off-bottom suspension. The HZH criterion is of a relative value only as it depends on particle size, density differences (particle/liquid), particle shape and solids concentration, and is thus applicable for scale-up under the conditions of geometric and hydrodynamic similarity.

Chudacek [98] has arrived at the constant tip speed criterion of a type of Eq. (6) with the help of an energy balance assuming geometric similarity and low solids concentration.

Another criterion results from a power consumed for homogenization for constant geometric ratio d/D:

$$P/V \sim N_p \varrho N^3 d^2 \tag{7}$$

While maintaining a constant impeller tip speed (Nd) and power number N_p we obtain for geometrically similar reactors that the power per unit volume on scale-up of homogeneous suspensions is inversely proportional to the linear scale-up ratio (D_2/D_1) or

$$P/V \sim 1/D \tag{8}$$

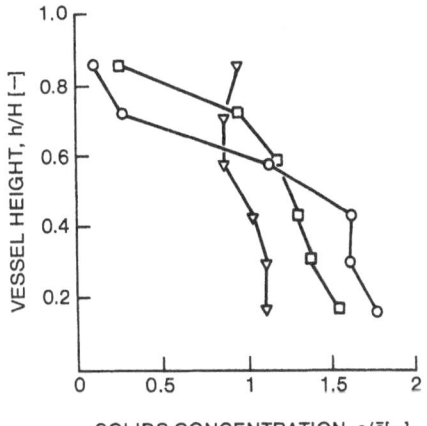

Fig. 12. Concentration profiles in the 0.48 m (D) reactor. Sand in water, $d_p = 157\ \mu m$, $\varphi = 9\%$ (v/v), rpm (min^{-1}): O — 234, ⌀ — 292, △ — 467 (adapted from Ref. [90])

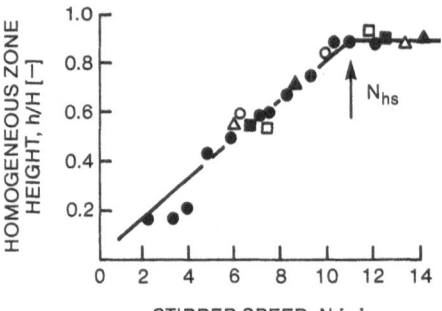

Fig. 13. Effect of stirrer speed on reactor homogeneity. Reactor D = 0.24 m, glass beads in water, d_p = 200 μm, $\varphi\%$ (v/v): $\bigcirc$ — 3, $\bullet$ — 7, $\square$ — 10, $\blacksquare$ — 20, $\triangle$ — 30, $\blacktriangle$ — 40 (adapted from Ref. [90])

Buurman et al. [90] also observed that in large-scale reactors concentration gradients develop as a function of agitator speed. From Fig. 12 it can be concluded that in the lower part of the vessel a zone of a relative high solids concentration is formed, where the concentration is approximately constant (homogeneous zone as used in this section). On the top of this zone the solids concentration decreases rapidly. With increasing stirrer speed, the height of the homogeneous zone will increase proportionally until a maximum value is attained of approximately 10% of the vessel diameter bellow the liquid level (Fig. 13). The solid concentration has no noticeable influence on the height of this zone. Data of Buurman et al. [90] are at variance with those of Einenkel [100] (compare Figs. 11 and 13), especially in terms of solid concentration effect. Also note an absence of N_{cs} break at low N in Fig. 13. Obviously more refined data are needed.

Buurman et al. [90] correlated their experimental data by a modified Froude number and found that the following inequality holds for a homogeneous suspension:

$$\frac{\varrho N_{hs} d^2}{g \, \Delta\varrho \, d_p} \frac{d_p^{0.45}}{d} \geq 20 \quad \text{or}$$

$$\frac{\varrho N_{hs}^2 d^{1.55}}{g \, \Delta\varrho \, d_p^{0.55}} \geq 20 \tag{9}$$

which implies that $N_{hs}^2 d^{1.55} \propto$ constant (or $N_{hs} d^{0.78} \propto$ constant). Note that product of N and d should be constant under the assumption of the largest eddies of approximately the impeller size (responsible for the homogeneity) as derived theoretically (Eq. (6)). In practice, Eq. (9) is close to the complete suspension relation (Eq. (1)). Obviously the eddies of the inertial sub-range also make a contribution to the homogeneity of the suspension. This means, that on a large scale, homogeneous suspension is attained at a lower specific power input (P/V) than on a small scale. Therefore, a constant P/V is not a suitable criterion for suspension homogenization (Eq. (8)).

Helical ribbon agitators are of interest because of their high homogenization efficiency at low agitator speeds. The reactor shape should be that of standard flat-bottom tank, which is easy to manufacture. Rieger et al. [101] measured homogenization (mixing) time of viscous liquids using standard decolorization method. No

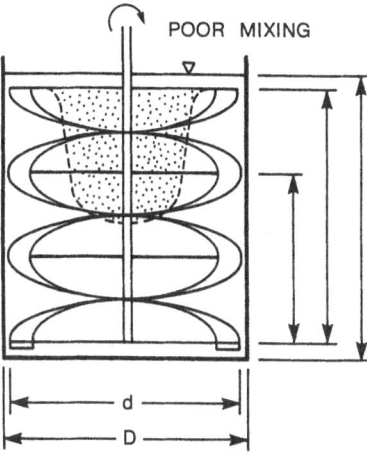

POOR MIXING

Fig. 14. Schematics of helical ribbon agitator with poorly mixed region (adapted from Ref. [101])

suspension homogenization data are available. Greater ribbon width and more blades than the standard geometry (2 blades) are suggested to achieve a better homogenization. Relatively poorly mixed region has been identified around the vessel axis (Fig. 14). This situation may not be favorable for an overlay aeration. Very gentle (low shear) mixing together with efficient homogenization and heat transfer (due to a minimum clearance between the agitator blade and the vessel wall) makes this agitator very suitable for mammalian cell application (microcarrier culture). In fact, Tolbert et al. [67] have successfully used a helix with a small pitch (sail type).

4.2.3 Significance

The above considerations are of importance for certain suspension cultures because the suspension homogeneity is of more significance than mass transfer. The usefulness of the uniformity should be appreciated in a high-density culture, such of maintenance bioreactor MBR. In this specific case, however, the homogeneity is achieved via intensive means of nutrient supply [67]. Moderately dense cultures of hybridomas, cells immobilized in microcapsules, entrapped in beads or in porous support material, and anchorage-dependent cells on microcarriers represent situations where the suspension criteria are of importance. Any gradients created within a suspension may impose a metabolic stress on cells in terms of lack of nutrients. It is thus imperative to control a degree of suspension. It may be, however, enough to eliminate sediments or deposits on the reactor bottom through an employment of unsuspended solids criteria. Bigger impeller diameter and small mixer clearance are obvious solutions. Better reactor geometry (bubble column with a draft tube, tapered or contoured bottom, etc.) offer other possibilities. In any case, N_{cs} can be established visually, if possible.

Homogeneous suspension requires more energy to achieve this status as compared to the elimination of sediments. Cells undergoing a cyclic exposure to different physical environment in a reactor experiencing zoning because of inhomogencity, may be induced to synthesize stress proteins or a certain differentiation state as a specific response to a given stimuli. Since the stress protein synthesis usually proceeds at the expense of other protein synthesis, the targeted product may not be thus produced in an optimal manner (Sect. 2.2). A constant tip speed (N_{hs} used for impeller speed) and/or power input are suggested for a scale-up of homogeneous suspensions, or possibly Eq. (9).

There is no single observation on the effect of cell deposits or reactor homogeneity on the product expression or on stress protein(s) synthesis to substantiate the significance of suspension criteria. However, there is no doubt that they are of importance in certain types of cell culture methods (high density) and reactor design.

4.3 Cellular Microenvironment Characterization

The previous chapter dealt with the status of external (bulk) cellular environment of relatively nondense culture. In this chapter the microenvironment is understood as a status of adjoining extracellular environment in a layer of cells attached to a support, in microcapsules, gel beads and other membrane configurations, all in terms of concentration profiles of nutrients, oxygen, products, etc. and pH, redox

Table 3. Summary of microenvironment homogeneity

Reactor type	Microenvironment type	Fig.	Ref.
Stirred tank:			
Suspension culture	Microscopically homogeneous		41)
Microcarriers			
Monolayer	Microscopically homogeneous		14)
Multilayer and bridging	Non-homogeneous		102)
Porous support	Microscopically homogeneous		54)
Microcapsules	Macroscopically homogeneous to non-homogeneous	15	55)
Column reactor:			
Stack or disc plate			
Monolayer	Microscopically homogeneous		44–45)
Fluidized bed porous support	(see stirred tank as above)		54)
Packed bed			
Glass spheres	Microscopically homogeneous		52)
Ceramic support	Microscopically homogeneous		51)
Membrane reactor:			
Hollow fibers			
extracapillary	Non-homogeneous		47)
lumen	Macroscopically homogeneous to microscopically homogeneous (cross-flow)		103)
Maintenance reactor (MBR)			67)
No support	Non-homogeneous		
Porous support	Microscopically homogeneous to non-homogeneous		
Microcarriers	Microscopically homogeneous to non-homogeneous		
Microcapsules	(see stirred tank as above)		55)
Entrappment	Macroscopically homogeneous to non-homogeneous (high density)	15	68)

and of other parameters of physical environment. It is important to realize what kind of profiles can be obtained with different cultivation methods and reactor configurations as growth of mammalian cells is restricted to a narrow range of environmental conditions and strict control of these greatly enhances growth. A summary of microenvironment homogeneity encountered with mammalian cells is in Table 3. In this table microscopically homogeneous environment is defined as that providing quite uniform level of microenvironment unless compromised by heavy sediments at the reactor bottom or by suspension nonhomogeneity. Note that such definition is not identical with a homogeneity due to micromixing at a molecular level [57]. In this category we consider also cells attached in the form of rather uniform monolayer. Macroscopically homogeneous environment provides a well defined layer of biomaterial controlled by capsule, bead size, or lumen (capillary) diameter. Of course, such environment is likely to be microscopically nonhomogeneous, depending on layer thickness, cell loading and their bioactivity. To eliminate the potential for nutrient and oxygen gradients along the length of fibers in a hollow fiber cartridge several units can be manifolded together in a parallel flow mode. The nonhomogeneous type covers situations difficult to control and resulting in substantial chemical and physical gradients.

Gradients, however, could be desirable from the production viewpoint. Thus axial flow along a plant cell layer entrapped between two membrane sheets evokes substantial nutrient concentration gradient (spatial heterogeneity) due to the diffusional limitation and consequently altered cell differentiation and biosynthetic pattern [104]. Such critical experiments have not yet been carried out with mammalian cells. However, it has been shown that the lumen immobilization of mammalian cells in a hollow fiber reactor is preferable for growth as compared to the extracapillary space [103]. Setec, Inc. [105] has recently introduced a hollow fiber cartridge with 100 μm effective culture thickness to eliminate nutrient gradients. The data of Schulz et al. [102] are in favor of a defined culture thickness. The specific productivity (in units per cell and time) of human β-interferon by mouse L-cells attached to microcarriers was highest for a monolayer culture and decreased significantly by the factor of 10 when cells formed trilayer on microcarriers. Any generalization would be premature as more data on different cell lines and products are needed.

The degree of packing (cell loading) and cells activity may also affect the degree of microenvironment homogeneity. For less dense environment of cells (e.g. in capsules) at an early phase of cultivation, there is a possibility of good degree of internal mixing inside the capsules and gradients will not be that pronounced. For a high-density culture, low external nutrient concentration and large capsule size gradients may be leading to a dead cell core zone (Fig. 15). The maximum physical packing density N_{max} of cells is determined by a cell volume. Pirt [106] derived a formula to calculate N_{max}:

$$N_{max} = 10^{12}/v \tag{10}$$

where v is cell volume in μm^3. Thus for a mammalian cell having 20 μm in diameter the maximum packing density is 2.4×10^8 cells per ml. The amount of cells of a given configuration is possible to support in a viable state, however, depends on their biological activity, external nutrient concentration and particularly on capsule size (thickness of culture between two adjoining membranes).

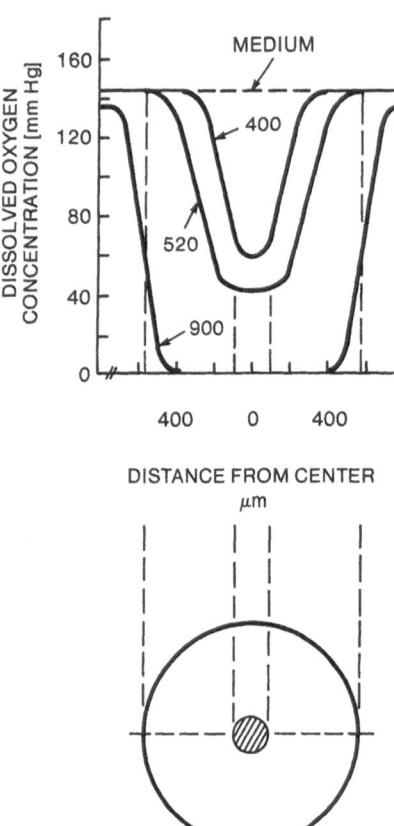

Fig. 15. Schematic representation of dissolved oxygen gradients within microcapsules packed with depleted zone for 1040 μm diameter capsule

4.4 Mixing Stress Characterization

Mammalian cells are subjected to a mechanical stress in any kind of bioreactor configuration, the most typical being a hydrodynamic stress during the mixing of a cell suspension. The hydrodynamic stress has two components, the first one due to viscous dissipation, predominating in a laminar flow, the second one due to fluctuating velocity vector in a turbulent regime. The hydrodynamic stress and some other stress effects will be discussed in the following.

4.4.1 Viscous Stress

The magnitude of shear stress in laminar regime is a function of fluid viscosity relative to the organism. Thus shear stress is a product of fluid viscosity (μ) and shear rate (γ) (velocity gradient):

$$\tau = \mu\gamma = -\mu \, dv/dy \tag{11}$$

Result of steady and time-independent stresses is a cell elongation (extension), followed by fragmentation (rupture). There are numerous examples in literature to

(1) Couette viscometer (inner cylinder rotating)

$$\tau = \frac{2\mu\, R_1^2 R_2^2}{r^2(R_2^2 - R_1^2)} \qquad (\text{dynes cm}^{-2})$$

Ω = angular velocity (rad s^{-1})
μ = viscosity (g cm^{-1} s^{-1})
R_1, R_2, r (cm)

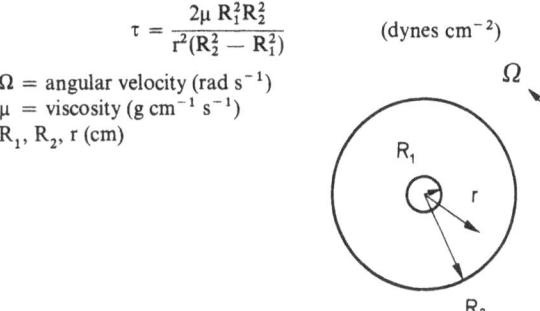

(2) Cone-and-plate viscometer

$$\tau = \frac{T}{2/3\ \pi r^3\ \sin \alpha} \qquad (\text{dynes cm}^{-2})$$

T = torque (dynes cm^{-1})

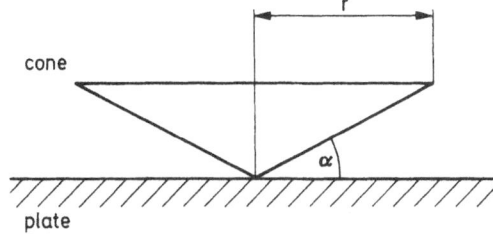

Fig. 16. Shear stress in laminar flow

document this type of mechanism: for red blood cells see Evans and Skalak [107]. In laboratory, well defined laminar flow regime with predominant viscous flow dissipation and minimal collisions and Reynolds stresses (see later) [108] can be generated in cone-and-plate or Couette viscometers. None of these standard equipments can be used for culturing of cells. Schürch et al [109 a] reported on specially constructed Couette device/bioreactor for a controlled shearing of hybridoma cells during a growth.

Laminar viscous regime only prevails at low values of the Reynolds number when the viscous forces are sufficiently large to damp out any small disturbances in the flow pattern. Thus well defined laminar flow as achieved in some viscometers has some merit in fundamental studies. The corresponding design (calculation) equations for two types of viscometers are in Fig. 16.

Laminar regime is also attained in some practical mixing devices, particularly in those with prevailing axial type of flow (Table 2). To determine whether a stirrer operates in a laminar regime one has to use a stirrer power characteristics in terms of the dimensionless quantities such as power number N_p and impeller Reynolds number Re. Such characteristics also provide a fluid zone division into laminar, transitional and turbulent regimes. The departure from laminar regime usually occurs between Re = 100 to 10,000. However, such correlations are available only for standard stirrer geometries [57].

4.4.2 Turbulent Stress

In a turbulent regime cells experience bursts of high and low shear stresses because of time and space fluctuations of liquid velocity. As a result, cells exposed in a turbulent flow experience several forces (Reynolds stresses), unlike those of steady time-independent ones in laminar field. The most significant may be as follows: a) Time-averaged shear forces and b) time-dependent shear forces. In practice, since the details of the turbulent motion are difficult to observe, most measurements of the fluid velocity will involve only time-averaged velocities. The time-averaged (and time-independent) forces can be evaluated at the maximum and mean. The maximum time-independent shear rate at the impeller tip can be estimated as follows [109]:

$$\gamma = N(1 + 5.3\beta)^{1/\beta} \, Re^{1/(\beta+1)} \tag{12}$$

where N is impeller speed. This correlation agrees well with the boundary layer theory. The mean shear rate in the bulk fluid can be estimated using the correlation of Calderbank and Moo-Young [110]:

$$\gamma = kN \tag{13}$$

Shear stress is then evaluated for low values of particle Reynolds number $Re_p \, (= \mu d_p/\nu)$ using Stokes flow

$$\tau = 3\mu\gamma/2 \qquad Re_p < 1 \tag{14}$$

For the Reynolds number greater than one

$$\tau = c_D \varrho u^2/2 \qquad 1 < Re_p < 100 \tag{15}$$

where

$$c_D = 18/(Re_p^{3/5}) \tag{16}$$

More fundamental description of flow results from fluid mechanical approach, using basic equations of motion, continuity, energy balance, etc. Thus from a force balance [111] it appears that the total time-averaged shear stress has two components, only one of them associated with turbulent phenomena:

$$\tau = \tau_v + \tau_t = u \frac{\partial u_m}{\partial y} + \varrho \overline{u'v'} \tag{17}$$

where τ_v stands for momentum transfer by microscopic mixing (viscous dissipation) and τ_t (or τ_r) is due to turbulent inertial (macroscopic) mixing and is called the Reynolds stress. The symbol u_m represents the space-averaged velocity and $\partial u_m/\partial y$

is the velocity gradient along the direction y. The ratio of the turbulent and viscous contributions to the stress can be estimated as

$$\frac{\tau_t}{\tau_v} = \frac{\varrho \overline{u'v'}}{u\, \partial u_m/\partial y} = \frac{\varrho u_{mm}^2}{u u_{mm}/D} = \frac{u_{mm}D}{v} \tag{18}$$

the Reynolds number characterizing the mean flow. The turbulent fluctuations u' and v' (streamwise and radial components) have been taken to be roughly proportional to the mean relative velocity, particle to fluid veclocity u_{mm} and the length over which the mean velocity varies significantly has been assumed to be comparable to the characteristic vessel diameter, D. Thus one may expect that turbulent stresses will be dominant in a flow with high Reynolds numbers, except very near the vessel wall.

From an energy balance consideration [111] it follows that the turbulence energy, far larger then the viscous dissipation, can be described as a mechanical energy, corresponding to time-dependent macroscopic velocity and pressure fluctuation in a vessel. Exact formulation of energy or force balances with time-dependent variables, however, does not provide solutions to a real situation. To meet these ends the only practical approach involves dimensional analysis, guided by certain physical simplifications.

Kolmogoroff theory of isotropic turbulence offers a convenient physical picture, based on energy transfer from a mixer through progressively smaller scales of the turbulence (eddies) in a process denoted as energy cascade [112]. In a flow situation with low Reynolds numbers, a perturbation due to mixing (largest possible disturbance is that of mixer scale) is eroded away by viscous dissipation. For high Reynolds numbers perturbations will result in a turbulency with prevailing Reynolds stresses. Kolmogoroff postulated that this occurs in an inertial sub-range of microscale eddies, for practical purposes the situation occurring in the impeller region. The length of eddies and fluctuating velocities of turbulence are dependent on the local energy dissipation. The average energy dissipation rate per unit mass can be estimated from the power number N_p

$$\varepsilon = \frac{N_p N^3 d^5}{V} \tag{19}$$

Actual energy dissipation, however, varies with location in a real vessel. Thus fluctuating velocity component of turbulence will also vary with the vessel location as derived on the basis of dimensional analysis (in inertial subrange):

$$u_f' \propto (\varepsilon d_p)^{1/3} \tag{20}$$

The inertial Reynolds stress forces, predominating in the stirrer region, can be then estimated by

$$\tau_r = C\varrho u_f^2 = C\varrho(\varepsilon d_p)^{2/3} \tag{21}$$

where C lies between 0.7 and 2 [113].

The viscous stresses exerted on the surface of a cell, predominantly in the bulk fluid, can be assessed similarly. Again, by dimensional reasoning

$$u_f' \propto d(\varepsilon_j/v)^{1/2} \tag{22}$$

and

$$\tau_v \propto \mu(\varepsilon_j/v)^{1/2} \tag{23}$$

and independent of cell size d_p. It can be shown that

$$\tau_v \propto du'/dx \tag{24}$$

a result analogous to a simple shear flow (Eq. (11)). It means that local velocity gradients may be represented by $(\varepsilon_j/v)^{1/2}$. For practical evaluation of viscous stresses the following equation can be used [114]:

$$\tau_v = 0.73\mu(\varepsilon_j/v)^{1/2} \tag{25}$$

The proportion of τ_r and τ_v can be then evaluated for respective zones of stirred tank provided it is divided into regions such as impeller tip zone, impeller zone and bulk [115]. Note, however, that in this analysis relative velocity fluctuations and local velocity fluctuations were approximated by relative particle to fluid velocity and that the isotropic assumption applied throughout [116]. An isotropic turbulence may, in fact, occur in reality with consequences such as nonuniform forces exerted on a cell surface.

As indicated above, the kinetic energy of turbulence is transported from large eddies to progressively smaller eddies in an energy cascade [111, 117]. The smallest eddies, which may perform a mechanical work on surface of cells, are defined as Kolmogoroff micro-scale eddies on the basis of dimensional analysis:

$$\eta = (v^3/\varepsilon)^{1/4} \tag{26}$$

where η is the characteristic length of micro-scale eddy. The Kolmogoroff micro-scale in stirred reactors is on the order of 25 to 200 μm [111, 118]. The smallest micro-scales are in the vicinity of the impeller and rapidly increase in size with increasing distance away from the impeller region.

For assessing the effects on cells suspended in a flow field the size of cells relative to micro-scale eddies is important. If the scale of eddies of the smallest turbulence is sufficiently larger than the cell size, cells will follow the local flow pattern (a), (Fig. 17) and move at the local liquid velocity. No shear damage is expected to cells as velocity differences between eddy streamlines are small [120], although particles can also rotate relative to the fluid (b, c). In addition to it, cells can move at a different particle-fluid relative velocity both in magnitude and direction (d). If the turbulent eddies are of the same order of magnitude as the cell size or less turbulent stresses may have some detrimental effect on cell activity. In that situation cells will experience fluctuating velocities on their surface even though the particle is in general moving with the fluid (e). Very small eddies will cause cells movement to

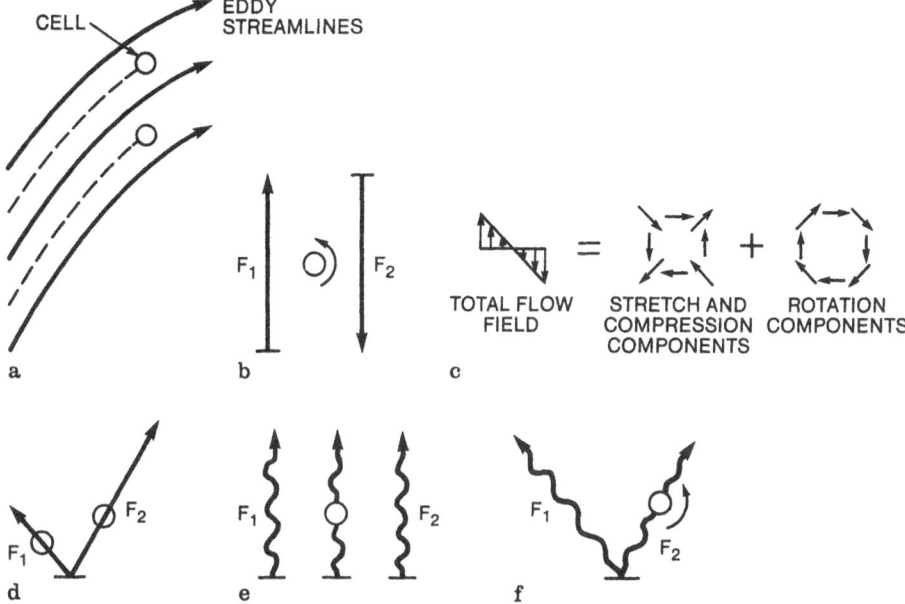

Fig. 17a–f. Schematics of particle-eddy turbulence (velocity) interactions (adapted from Refs. [119] and [93]). Explanations of **a, b, d–f** in the text: **c** decomposition of linear shear field into stretch and rotation components. Note that only the stretching components act to disrupt a cell

uncouple from large-scale eddy flows. Finally, both particle and fluid velocities will vary spatially throughout the vessel and particles will themselves move turbulently with random fluctuations in any direction and they will also rotate (f).

Since the measurement of this complex velocity in specific regions of the vessel has proved difficult some macroscopic estimates are possible. The inertial forces (Reynolds stress tensor) can be estimated by Eq. (21) and viscous stresses, which represent only a fraction of total stresses, through Eq. (25).

Rosenberg [121] carried the above calculations for plant cells grown in a mixed vessel. The bulk viscous stresses for his conditions were evaluated on the order of 0.2 to 13 dynes cm^{-2}. Such level of stress is significantly lower than that required to inhibit growth and cell division in a laminar flow field (50 to 100 dynes cm^{-2}).

Based on analogy with flocs disintegration in a turbulent field, one can assume that the physical mechanism of cell damage may involve bulgy deformations, followed by a loss of cell membrane integrity (perforation) [113]. Rosenberg [121] obtained a preliminary evidence for this type of mechanism in plant cells.

4.4.3 Stress at Boundary Layer and Collisions

An additional consideration in an agitated bioreactor is that suspended cells experience cell-cell and/or cell-impeller collisions that may be damaging [119]. In a typical reactor a cell will collide with impeller, baffles, walls, internal probes and other cells. The damage done (severity) to the cells will be a function of collision frequency and the amount of inertial energy transmitted to a cell at the collision event. Both terms have been defined by Cherry and Papoutsakis [119], based on work of Hinze [122] and assuming that cells' velocity is equal to the characteristic velocity derived from Kolmogoroff theory. The turbulent collision severity (TCS) is then:

$$TCS = (\varepsilon v)^{3/4} \left(\frac{\pi^2 \varrho_c \varphi d_c^4}{36} \right) \tag{27}$$

The interaction can be either direct physical contact between collision partners or the modification of the local shear field around each collision partner as they come to a close proximity. In order for cells to actually make physical contact, the inertial forces driving collision partners towards each other must be greater than the forces necessary to drive the liquid out from between them [123]. The above analysis, however, assumes that the collision event is inelastic. In reality, the amount of energy transmitted to a cell at collision will depend on the kinetic energy of cells and elasticity of the collision. Mammalian cells, having an elastic membrane, will deform during the collision and experience thus less damaging force as compared to a rigid surface. A deformation beyond a certain point in both laminar and turbulent regimes may be damaging to cells and their viability may be compromised. Red blood cells, having a simple cytoskeleton, have evolved into bodies capable of considerable extension and stretching during a flow in capillaries or at an impact without compromising their viability [107]. Other mammalian cells with more complex cytoskeleton are more rigid. The elasticity of the event has not yet been brought into a theoretical treatment of collision damage.

The cell-impeller collision is assumed to occur by a process of interception in which cells passing within one diameter of the central streamline actually collide with the impeller face. The real physical mechanism could be same as implied in cell-cell collision, i.e. a contact or hydrodynamic film modification. Rosenberg [121] has modified Cherry and Papoutsakis's expression for cell-impeller collision frequency by considering the impeller and a cell as collision partners (incorporating cell volume fraction rather than the reactor volume occupied by sweeping impeller) and arrived at:

$$N_i = \frac{18 n_b d^2 N \varphi}{8 d_c^2} \tag{28}$$

and at cell-impeller collision severity (ICS):

$$ICS = 1.71 \varrho_c n_b N^3 \varphi d_c d^4 \tag{29}$$

Rosenberg [121] calculated both cell-cell and cell-impeller collision frequencies and severities for his plant cell culture having a mean cell aggregate size of 180 µm. The kinetic energy of cell-impeller collisions was several orders of magnitude greater than the one of cell-cell collisions due to the high velocity of the impeller relative to the motion of a cell. Furthermore, the frequency of cell-impeller collisions was also greater than cell-cell collisions. Thus, cell-impeller collisions are clearly more significant than cell-cell collisions. The relative contributions of hydrodynamic shear and collisions have not yet been quantified.

A rather special interaction occurs with cells growing attached to microcarriers. Based on experimental data with an addition of inert beads of the same size and buoyancy Crougham et al. [124] identified the microcarrier-microcarrier interaction (collision) as a second order event in the microcarrier concentration. Perhaps, cells are damaged when microcarriers collide between themselves. In addition to that the same authors also postulated a hydrodynamic damage of cells due to the eddy-microcarrier interaction (first order in microcarrier concentration).

4.4.4 Stress due to Aeration and Air-Liquid Interfacial Phenomena

Direct sparging of gas has been shown to be detrimental to cell viability and proliferation [125–127]. It is suspected that the deleterious effect of sparging is associated with the surface energy of the air bubble [128]. The surface forces due to surface tension can presumably stretch the cellular lipid bilayer and result in the cell damage. Aunins et al. [129] suggested a formula for minimal and maximal local shear rate due to bubble aeration. The minimum local rate is:

$$\gamma_{min} = 1/2v_t N_b^{0.33} \tag{30}$$

The upper limit in the trailing vortices of rising bubbles is as follows:

$$\gamma_{max} = 1/2(v_t^3/vR_b)^{1/2} \tag{31}$$

where v_t is bubble terminal velocity and R_b bubble radius. They estimated γ_{min} and γ_{max} to be 18 s^{-1} and 2900 s^{-1}, respectively. Mercer [130] gives bubble velocities in an air-lift bioreactor in the order of 25 to 50 cm s^{-1} which results in a local maximum shear stress in the order of 10 to 100 dynes cm^{-2}. Clark et al. [131] reported on maximal shear stress at wall of an air-lift column to be 120 dynes cm^{-2}. Such values are high enough to cause a cell damage. The effect seems to be limited only to the region of bubble disengagement while a bubble is formed [132].

The damaging effect of sparging can be minimized by using large air bubbles or low gassing rate in mechanically-agitated vessels. Serum is believed to stabilize cell-liquid interfaces compared to serum-free medium, although promoting foaming. Also, some other polymers (e.g. PEG) protect cells because of their ability to form stable foams [133]. An intermittent sparging of pure oxygen is also used. Continuous sparging of air is not suitable for microcarrier culture primarily due to an accumulation of beads at the air-liquid interface (foam).

The role of surface shear in cell physiology has not yet been sufficiently elucidated. Besides cells damage at the air-liquid interface product or nutrient denaturation should be also considered.

Extracellular protein products are likely to be also affected at air-liquid interfaces. The alternation of tertiary protein structure denoted as denaturation may result from an exposure at the air-liquid interface [133]. It is assumed that proteins form a spread film with surface pressure causing unfolding. As the denaturation seems to be reduced in the presence of surface-active substances, this fact may be brought into a consideration during the downstream processing from serum-containing media.

4.4.5 Physiological and Morphological Effects of Mixing Stress

Physiological responses to any kind of stimuli have been mentioned in Sect. 2. It is assumed that some portion of the stimulus-response cascade is shared in case of a mechanical stimuli [134]. Two last steps of the cascade represent responses which may occur as a result of subcritical (i.e. non-damaging) level of stress, namely gene activation, protein expression and numerous differentiation responses. The subcritical levels of stress may thus have a relevancy to practical situations. A stimulation of urokinase production by relatively low shear stress in human embryonic kidney cells may serve as convenient example [134]. The employment of elevated state of metabolism or of a particular differentiated state through moderate stress level has advanced too little.

The above mixing stress discussion can be generalized little further. The cell-cell contact in dense mammalian cultures (confluent cultures on support) may exert a form of stress by adjoining cells. A similar situation may exist in the maintenance bioreactor (MBR) or with encapsulated cells. Ben-Ze'ev [135] demonstrated that some cytoskeleton components undergo dramatic changes in a dense epithelium cell culture as compared to a nondense ones. Napolitano et al. [136] observed the appearance of previously undetected polypeptide, perhaps a novel stress protein. Extensive cell contact may constitute a stressful situation resulting in the increased production of a class of proteins, accompanied by a decrease of ordinary protein synthesis. Such mechanism may represent a way how a high density culture switches to a maintenance non-proliferating state, capable of sustained viability and of some biological function (expression of a desired class of proteins). The practical consequences are obvious.

5 Oxygen Transfer and Design

5.1 Oxygen Uptake Rates

Oxygen demand of mammalian cells is much less compared to the other microbial systems. Fleischaker and Sinskey [137] compiled a list of measured oxygen uptake rates of different human cells in a culture. The uptake falls of most cells falls in the range of 0.05 to 0.5 mmol l^{-1} h^{-1} oxygen (= 1.6 to 15 mg l^{-1} h^{-1}) at 10^6 cells per ml, one to two orders of magnitude less as compared to yeast cells. The numbers mentioned represent the maximum uptake rates. The actual uptake rates depend on dissolved oxygen concentration. Spier and Griffiths [128] reported a list of values of optimal oxygen concentration for growth of about ten cell lines. It appears that when medium used for growing cells is equilibrated with an oxygen/nitrogen mixture in which the partial pressure of oxygen is about 10 (range is to 5 to 20%) of the total pressure, the resulting concentration of dissolved oxygen is optimal for a wide range of cells in a variety of culture systems. The mentioned range probably reflects an evolutionary adaptation of cells in a particular tissue location.

5.2 Monolayer Aeration

A mathematical model of gas diffusion in a static cell culture represented by a monolayer (Roux flask, T-flask) has been developed by McLimans et al. [138] which relates the depth of the culture fluid to the oxygen available to cells. A critical maximum depth of 1.2 mm has been identified guaranteeing an adequate oxygen supply for moderately demanding cells.

Spier and Griffiths [128] estimated oxygen transfer rate in 5 mm deep static culture assuming oxygen saturation at the top of liquid and zero concentration at the monolayer. The value is $1.5 \, \mu g \, cm^{-2} \, h^{-1}$. The same authors, however, obtained experimentally about $17 \, \mu g \, cm^{-2} \, h^{-1}$ for oxygen transfer into a large static volume. The above numbers can serve as a guide to estimate the cell density limits of static cultures provided the oxygen uptake rates are known.

A roller bottle with cells attached to the internal plastic surface represents another form of monolayer culture with renewal of liquid surface. Conceptually similar is an aeration in a rotating tube partially filled with liquid. Moser [139] predicted the mass transfer coefficient (k_L) dependency on speed of rotation and Mukhopadhyay and Ghose [140] confirmed this dependency experimentally. These findings are in agreement with the surface renewal theory of Danckwerts [141]:

$$k_L \propto N^{1/2} \tag{32}$$

Phillips et al. [142] obtained experimentally k_L values of 6 to 75 cm h^{-1} at 0–100 rpm, Coppock and Meiklejohn [143] reported k_L in the range of 100 to 200 cm h^{-1} and Moser [144] $k_L a$ of 200 h^{-1} at 200 rpm.

5.3 Overlay Aeration

Overlay (overhead) or surface aeration represents mass transfer across relatively unbroken surface in mildly agitated culture and is generally less effective compared to a monolayer aeration. Empirically, one can propose a relationship between Sherwood (Sh_s) and impeller Reynolds (Re) numbers (neglecting other dimensionless groups) as suggested by Aunins et al. [129]:

$$Sh_s \left(= \frac{k_L H}{D_{02}} \right) = f(Re)^c = f \left(\frac{ND^2}{v} \right)^c \tag{33}$$

From Eq. (33) it can be shown that k_L is directly proportional to the impeller speed N, air flow rate across the interface (overlay air flow rate) and indirectly to the culture depth or volume. Fleischaker and Sinskey [137] found that $k_L a$ (a product of k_L and of air-liquid interfacial area a) can be correlated to N and V:

$$k_L a = kN/V^{2.05} \tag{34}$$

The air flow rate determines liquid mixing (velocity of circulating eddies) and consequently the depth to which fluid carries newly saturated liquid from the surface into

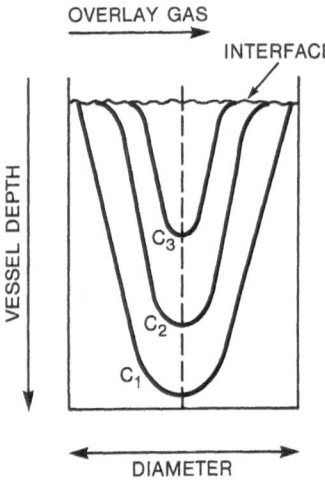

Fig. 18. Dissolved oxygen gradient at overlay aeration (adapted from Ref. [144a])

the bulk liquid. Obviously, the mixing pattern will then result in dissolved oxygen gradients [144a], (Fig. 18).

The exponent in Eq. (33) can vary between 2/3 to 3.0 depending on hydrodynamic conditions. Exponent $c = 0.75$ can be predicted on the basis of eddy diffusivity model of King (Chen et al. [145]) and applies for surfactant loaded systems, typical in mammalian cultures with protein(s) stabilized films. Aunins et al. [129] indeed experimentally confirmed the exponent to be 0.73. A high c (1.5) is relevant for situations where interface is in a state of transient diffusion, as represented by the surface renewal theory of Danckwerts [141]. One way to overcome the presence of surface films is to use a surface impeller in addition to an immersed mixer. Hu et al. [146] obtained $c = 3.0$ for a two disc impeller mixer, one serving as a surface impeller.

Table 4. Oxygen demand vs. surface aeration for 10^6 cells per ml culture

Culture volume (liter)	Head space area (cm²)	Oxygen supply by overlay aeration (µg h⁻¹)	1	Oxygen demand (µg h⁻¹) 2	3
(a) k_L (overlay aeration by air) = 4 cm h⁻¹					
1	100	3,000	1,000	2,000	3,000
10	500	15,000	10,000	20,000	30,000
100	2,500	75,000	100,000	200,000	300,000
			5	10	15
(b) k_L (overlay aeration by air) = 80 cm h⁻¹					
1	100	60,000	5,000	10,000	15,000
10	500	300,000	50,000	100,000	150,000
100	2,500	1,500,000	500,000	1,000,000	1,500,000

Liquid height (H)/vessel diameter (D) = 1;
Enclosed area represents oxygen-limited regime (adapted from Ref. [147])

Aunins et al. [129] tabulated coefficient c of Eq. (33) from available literature data for a free surface transfer in turbulent systems.

Katinger and Scheirer [147] presented measured k_L values for surface aeration in the range of 2 to 4 cm h^{-1}. Based on that they compared oxygen demand with oxygen supply (Table 4). It appears that for very low oxygen demands overlay aeration may become critical for vessels between 1 to 10 liters. Spier and Griffiths [128] reported k_L in the range of 2 to 80 cm h^{-1} for spinner flasks overlay. Using the upper values the Table 4 presents more favorable situations extending to higher oxygen demands. For cell densities higher than 10^6 cells per ml culture will be still oxygen-limited. An additional surface aerator [146] or oxygen enrichment is a solution at hand.

5.4 Pure Oxygen Aeration

Contrary to a general belief the pure oxygen aeration can be used to satisfy a culture demand. It is, however, advisable to supply it in short pulses via a dissolved oxygen control loop. The tuning of the control circuit should be in the way that there are no major differences between upper and lower levels of oxygen at on-off control. Low overall oxygen throughput rate usually does not cause any substantial foaming.

5.5 Air Sparging

Spier and Griffiths [128] and subsequently New Brunswick Scientific, Inc. [59] introduced a caged aeration. To eliminate the cell damage at the air-liquid interface liquid phase is separately aerated in an enclosed space (cage). The fine screen does not allow the cells on microcarriers to enter the aerated space. Air or intermittent pure oxygen aeration can be used [59]. Reported $k_L a$ is in the range of 1 to 13 h^{-1} (air) or 10 to 90 h^{-1} (pure oxygen) for a 5 liter vessel. Reuveny et al. [59] also reported values on their 16 and 75 liter bioreactors.

5.6 Special Oxygen Supply Systems

Oxygen can be supplied via a semi-permeable silicone tubing. Fleischaker and Sinskey [137] and later Aunins et al. [129] demonstrated its utility for smaller culture systems. An empirical relationship similar to Eq. (33) has been proposed. Also a scale-up procedure has been suggested. This type of aeration would not be a practical solution for large vessels as the tubing length would be excessive. However, based on their data with porous hydrophobic polypropylene fibers, Lehmann et al. [148] envisioned an application to a scale of 300 liters.

6 Bioreactor Scale-Up

In this chapter only special considerations will be presented inherent to mammalian culture bioreactor. It would be first advantageous to mention volumes of bioreactors typically encountered at production. The maximum volume reported for a suspension culture of hybridomas is about 8,000 liters for stirred bioreactor [149] and 100 liters for an air-lift bioreactor [150]. The bioreactor scale-up for these configurations should follow well-established rules of microbial technology [151] with a special attention to shear sensitivity of hybridoma cells. A microcarrier culture of 1,000 liters has been reported by Montagnon [152] and of 4,000 liters by Nilsson [153]. It is believed that the industry uses up to 10,000 liters. Verax Corporation [54] reported on a pilot production facility employing a porous carrier system in a fluidized reactor. Nilsson [153] reported on their reactor volume of 800 liters. Damon Biotech uses volumes between 20 to 50 liters with their encapsulation method. The volumes mentioned above are not excessive and scale-up factor is moderate. Nevertheless, because of many special processing features of mammalian cell cultures, the scale-up to such volumes is a challenging task.

In a suspension culture the availability of oxygen and mass transfer could be first factor to be considered. Direct sparging is not universally acceptable because of excessive foaming and possible shear effects at the interfaces. An intermittent air (oxygen) supply through a convenient controller (and timer) would minimize these effects. For an overlay aeration Eq. (34) may serve as a guideline for the effect of reactor volume and impeller rotation speed. A surface aerator, in addition to a regular impeller (propeller) would provide better oxygen supply capability. Perhaps, Eq. (33) with appropritate exponent c should be used for scale-up.

Next parameter to be considered at scale-up is cell's fragility. Impeller tip speed [3], representing the maximum time-averaged shear stress, is typically employed, primarily because of lack of understanding of complex phenomena occurring at mixing (Sect. 4.4). This approach fully neglects some other stresses, particularly those at collision and at air-liquid interfaces. Unless further advances are made in the understanding of mechanism of shear damage at a cellular level and of inactivation kinetics no substantial progress can be expected in the mammalian cell bioreactor scale-up area.

The last possible parameter at scale-up is the complete and/or homogeneous suspension criteria approach, applicable for dense suspensions or for porous carrier-immobilized cell systems. Based on discussion in Sect. 4.2 several scale-up criteria can be put forward. The consideration of unsuspended solids suspension (complete off-bottom suspension) may lead to a power/volume criterion (Eq. (8)) or to a particular product of N_{cs} and $d^{2/3}$ (Eq. (2)) or to a special tip speed criterion for vessels provided with draft tube or for loop reactors ($N_{cs}d$). Such criteria will guarantee a desirable micro environment believed to be necessary for a suitable cell physiology and product expression. The homogeneous suspension approach also leads to modified tip speed criteria, where in place of N a stirrer speed at a homogeneous suspension N_{hs} is used (Eqs. (6) and (9). The applicability of these criteria, developed for non-biological systems, will have to be tested in a mammalian cell culture. The first logical step would be to look into Eq. (2) to facilitate removal of sediments at reactor bottom.

A special scale-up procedure employed with membrane bioreactors or with ceramic matrix is that of linear enlargement and parallel multiplication of membrane units (e.g. a hollow fiber cartridge can be scaled-up to a certain size and later multiplied in number as employed by several companies in their production units). Obviously, there is an upper limit due to inherent gradients. However, such a limit represents many times more over the conventional suspension technology in terms of productivity.

7 Bioreactor Sterilization and Containment

Again, special considerations will be given in this chapter. Due to the extreme sensitivity of mammalian cell culture to any type of contamination exceedingly high requirements are used to ensure the bioreactor sterility. Usually recommended level of assurance of at least 10^{-6} probability of survival (non-sterility) for bioreactor sterilization [154] is not acceptable for mammalian cell bioreactors. As the media components are not sterilized together with a bioreactor an overkill approach can be employed for heat stable products providing assurance of to 10^{-12} probability of survival (fewer than one non-sterile unit per one billion units). The procedures how to guarantee such a level of sterility are outlined in already mentioned monograph [154] and a general framework is germane to GMP [155, 156].

Heat labile media are readily filtered, frequently several times, typically with 0.45 μm, less with 0.1 μm average pore size cartridges. The methodology for assuring filter integrity and sterility of products is available [157, 158]. It should be mentioned that the main reason why a multiple filtration should be utilized is the presence of mycoplasmas in serum, used as a necessary component in serum media. Even such treatment does not remove mycoplasmas completely although claimed by manufacturers (e.g. HyClone Labs., Inc.). In-line filters for handling air and liquid streams to and from a bioreactor are important components of the physical containment system in case of rDNA organisms. The integrity of filters of any stream should be carefully checked via a validation procedure. Design requirements of large containment systems are briefly discussed in [159]. They are quite complex and capable to handle both normal and emergency releases of rDNA organisms.

8 Bioreactor Instrumentation and Control

Control strategies for optimizing mammalian cells metabolism have not been developed adequately [1]. It is due to the fact that low growth and production rates in typical mammalian cultures make the task of monitoring and control very difficult. The following approaches may develop into a valuable tool providing an estimate of the metabolic status of cells:
1) Oxygen uptake and related strategies;
2) glucose and glutamine feeed strategy;
3) internal redox status (NAD(H)) of cells;
4) in vivo nuclear magnetic resonance (NMR) monitoring.
Oxygen uptake by cells, measured through a balancing, is very easily obtained. Coupled with the lactate production rate it can be used for an on-line estimation of cell concentration [160] and subsequently for glucose feed adjustment. Charles River Biotechnical Services, Inc. employed a similar strategy for glucose feed in their ceramic matrix-based bioreactor [161]. High density cultures, as in the case of the latter example, have high chances of reliable control because of higher oxygen uptake across the bioreactor.

Glacken et al. [160] developed a control strategy to reduce production of lactate and ammonia via controlled addition of glucose and glutamine. Both lactate and ammonia may cause an intoxication of a mammalian culture.

On-line monitoring of cells metabolic status, particularly sensitive to oxygen or substrate supply, is possible through a fluorescence probe detecting an internal concentration of NAD(H) [162]. A feeding schedule can be determined to maintain a particular redox status (ratio of reduced to oxidized coenzyme).

The last prospective method is in vivo non-invasive monitoring of culture status via NMR [163]. Drury and Dale [164] constructed a special hollow fiber growth chamber which can be fitted into a standard NMR spectrometer. The method, as it stands, is however, limited to a high density cultures. The latest improvements in the resolution of NMR will allow for a single cell imaging [165] or for gradients imaging in high density cultures, such as in capsules, beads, hollow fiber reactors, etc.

9 Conclusions and Outlook

In the following some areas will be discussed which need further attention to facilitate a progress in practical mammalian cell culture systems:

1) Cell biology progress in signal processing, that is how any type of input is processed by a cell, may be useful for our understanding of cell functions and eventually may be employed for some production goals.

2) Suspension criteria presented in Sect. 4.2 should be applied in specific cases and their utility tested. The status of microenvironment is considered to be a key for an understanding for many types of cell culture and of different protein expression systems.

3) The characterization of most adjacent to cell microenvironment status is considered as an urgent task. Perhaps, employment of micro-versions of dissolved oxygen, pH and redox potential (NAD(H)) probes would be desirable in conjunction with different high density culture configurations.

4) Most compelling is a problem of differentiation between different shear stress components, their mechanism of action and biological responses. Eventually, controlled moderate shear should be employed towards a production increase. Shear at the air-liquid interface has not obtained a deserved attention, particularly in terms of mechanisms involved.

5) Design of a gentle (nonshearing) mixing with satisfactory oxygen supply is open to exploitations. Some untested examples have been listed in Table 2.

6) Two areas of bioreactor scale-up deserve an attention. First, based on shear considerations, second, based on suspension criteria, particularly useful in microcarrier technology.

7) Increased attention to sterility control may lead to serum-free media (some other considerations may be even more important, e.g. downstream processing) to reduce a contamination risk by mycoplasmas and viral elements.

8) Bioreactor control has been very much a neglected area. Development of new probes and control strategies, preferably on cells status principles, would greatly improve our production capabilities.

10 Acknowledgement

The authors would like to thank Dr. R. K. Bajpai for constructive criticism leading to a great improvement of this article.

11 Nomenclature

a	global air-liquid interface
b	stirrer blade thickness
c	exponent in Eq. (33)
	solids concentration
$\bar{c}$	mean solids concentration
C	constant in Eq. (21)
C_D	drag coefficient
d	impeller diameter
d_c	cell size
d_p	particle size
D	(characteristic) vessel dimension (diameter)
D_{02}	oxygen diffusivity in medium
$F_{1,2}$	forces at flow
g	gravitational constant
h	homogeneous zone height
H	liquid height
HZH	homogeneous zone height (h/H, dimensionless)
ICS	cell-impeller collision severity
k_L	mass transfer (surface) coefficient
$k_L a$	volumetric mass transfer coefficient
$\bar{n}$	average fluid velocity at a level (Eq. 3)
n_b	number of impeller blades
N	impeller speed
N_b	bubble concentration
N_i	cell-impeller collision frequency
N_p	power number
N_{max}	maximum cell packing density
N_{cs}	impeller speed at complete solids suspension
N_{hs}	impeller speed at homogeneous suspension
P	power input
r	annulus radius
R_1	inner cylinder radius
R_2	outer cylinder radius
Re	impeller Reynolds number
Re_p	particle Reynolds number
R_b	bubble radius
Sh_s	surface Sherwood number
T	torque
TCS	turbulent collision severity

u	settling particle (cell) velocity
u_m	spatial mean velocity
u_{mm}	particle-fluid relative velocity
u'	fluctuating fluid velocity component (coordinate x)
u'_f	root mean square value of fluctuating velocity component of turbulence
$\bar{u}_{base}$	mean liquid velocity at the bottom
v	fluid velocity
	cell volume
v'	fluctuating fluid velocity component (coordinate y)
v_t	bubble terminal velocity
v_{sg}	superficial gas velocity
V	liquid volume
x	streamline flow component
y	radial flow component
α	cone-and-plate viscometer angle
β	coefficient in Eq. (12)
γ	shear rate
ε	energy dissipation rate
ε_j	local energy dissipation rate
η	characteristic length of micro-scale eddy (Kolmogoroff micro-scale)
μ	fluid viscosity
ν	kinematic viscosity
ϱ	fluid density
$\Delta\varrho$	phase density difference
ϱ_c	cell density
τ	shear stress
τ_r	Reynolds stress
τ_t	turbulent stress
τ_v	viscous stress
φ	particle (cell) volume fraction

12 Note Added in Proof

Sprague et al. [166] reported on details of shear stress-related receptor internalization and degradation (cf. Ref. [9]). Lambe and Walker [167] presented a survey on reactor requirements for animal cells, Griffiths et al. [168] compared reactor possibilities for anchorage-dependent cells and Fleischaker [169] reviewed the current status of micro-carrier cell culture. A comparison of different modes of operation (batch, fed-batch, continuous, perfusion) in terms of monoclonal antibody production ($mg^{-1} d^{-1}$) has been carried out by Macmillan et al. [170]. The very first data on the microcarrier (Cytodex 3) concentration profile within the 20 and 150 liter bioreactors (plot is similar to that of Fig. 12) as affected by stirrer speed (tip speed) have been presented by Lehmann et al. [171]. Tip speeds 8–10 cm s^{-1} provided almost uniform profile, without substantial microcarrier deposit at the vessel bottom. In the same paper authors described in detail their membrane-stirrer system, enabling simultaneous aeration and perfusion. Wichterle [172] developed a theoretical model for suspension of solid particles based on the comparison of terminal particle velocity and the characteristic liquid velocity at the bottom. The model explains adequately empirical observations of other investigators. Smith et al. [173] presented detailed kinetic investigation on the shear effect on hybridomas, using cell viability and LDH release as criteria. Robertson and Ulbrecht [174] derived an expression for the maximal shear rate in a mechanically mixed vessel and confirmed it experimentally. The maximal rate at the impeller tip is proportional to the agitator diameter and to the stirrer speed (with 3/2 power). McQueen et al. [175] estimated that the Kolmogoroff micro-scale, characterizing the size of energy-dissipating eddies, is close to hybridoma cell size. Based on experimental data of turbulent flow in capillaries, the significant loss of hybridoma viability occured when the Kolmogoroff micro-scale reached a value somewhat smaller than the cell size. Hirschel and Gruenberg [176] presented details of Endotronic's hollow fiber technology with description of their control strategy. Feder [177] disclosed few details on Monsanto's static maintenance bioreactor (now denoted as perfusion maintenance reactor). Lower media requirements (incl. serum) seem to be one major advantage. No quantitative data were presented to support this statement. However, with the cell density increase the perfusion rates have to be disproportionally increased offsetting the advantages of this reactor concept. Thus 100-fold cell density increase results in 100-fold total protein produced and only 8-fold increase in the product specific activity. Defining the maintenance requirements of non-proliferating animal cells and maintenance reactor homogeneity require further attention, especially with the reactor scale-up.

13 References

1. Glacken MW, Fleischaker RJ, Sinskey AJ (1983) Ann. New York Acad. Sci. *413*: 355
2. Hu W-S, Dodge TC (1985) Biotech. Progr. *1*: 209
3. Arathoon WR, Birch JR (1986) Science *232*: 1390
4. Adamson SR, Schmidli B (1986) Can. J. Chem. Eng. *64*: 531
5. Guharay F (1984) J. Physiol. *352*: 685
6. Lausman JB, Hallam TJ, Rink TJ (1987) Nature *325*: 811
7. Salisbury JL, Condeelis JS, Satir P. In: Richter GW, Epstein, MA (eds) International review of experimental pathology, vol 24, Academic, New York, p. 1
8. Goldstein JL, Brown MS, Anderson RGW, Russell DW, Schneider WJ (1985) In: Palade GE, Alberts BM, Spudich JA (eds) Annual reviews of cell biology, vol 1, Annual Reviews, Palo Alto, p 1
9. Schwartz CJ, Nerem RM, Sprague EA, Levesque MJ, Steinbach BL (1986) Biorheology *23*: 195
10. Fulton AB (1984) The cytoskeleton. Cellular architecture and choreography, Chapman and Hall, New York
11. Darnell J, Lodish H, Baltimore D (1986) Molecular cell biology, Scientific American Books, New York
12. Langanger B, Moeresmans M, Daneels G, Sobieszek A, De Brabander M, De Mey J (1986) J. Cell Biol. *102*: 200

13. Cleveland DW (1982) Cell 28: 689
14. Welch WJ, Suhan JP (1986) J. Cell Biol. 103: 2035
15. Freshney RI (1987) Culture of animal cells. A manual of basic technique, 2nd ed., A. R. Liss, New York
16. Griffiths JB, Riley PA (1985) In: Spier RE, Griffiths JB (eds) Animal cell biotechnology, vol. 1, Academic, London, p 17
17. Ben-Ze'ev A, Farmer SR, Penman S (1980) Cell 21: 365
18. Quintanilla M, Brown K, Ramsden M, Balmain A (1986) Nature 322: 78
19. Kapp-Pierce J (1987) personal communication, Washington University Medical School, St. Louis
20. Barnes D (1987) Biotechniques 5: 534
21. Eshhar Z (1985) In: Springer TA (ed) Hybridoma technology in the bioscience and medicine, Plenum, New York, p 3
22. Varani J, Wass JA, Rao KMK (1983) J. Natl. Cancer Inst. 70: 805
23. Mely-Goubert B, Bellgran D (1981) J. Immunol. 127: 399
24. Gowingt LR, Tellam RL, Banyard MRC (1984) J. Cell Sci. 69: 137
25. Chen LB, Rosenberg S, Nodakavukaren KK, Walker ES, Shepherd EL, Steerle Jr GD (1985) In: Springer TA (ed) Hybridoma technology in the bioscience and medicine, Plenum, New York
26. Pasternak C, Elson EL (1985) J. Cell Biol. 10: 860
27. Mazur MT, Williamson JR (1977) J. Cell Biol. 75: 185
28. Alberts B, Bray D, Lewis J, Raff M, Roberts K, Watson JD (1983) Molecular biology of the cell, Garland, New York
29. Rosenberg MZ, Prokop A, Kargi F, Dunlop E (1988) In: Fiechter A (ed) Adv. biochem. engin./biotechnol., Springer, Berlin Heidelberg New York, submitted
30. Packard B (1986) Trends Biochem. Sci. 11: 154
31. Berridge MJ, Irvine RF (1984) Nature 312: 315
32. Bell RM (1986) Cell 45: 631
33. Lindquist S (1986) In: Richardson CC, Boyer PD, Dawid IB, Meister A (eds) Annual review of biochemistry, vol 55, Annual Reviews, Palo Alto, p 1151
34. Stathopoulos NA, Hellums JD (1985) Biotech. Bioeng. 27: 1021
35. Burdon RH (1986) Biochem. J. 240: 313
36. Pelham HRB (1986) Cell 46: 959
37. Taglicht D, Padan E, Oppenheim AB, Schuldiner S (1987) J. Bacteriol 169: 885
38. Parag HA, Raboy B, Kulka RG (1987) EMBO J. 6: 55
39. Spier RE, Fowler MW (1985) In: Moo-Young M (ed) Comprehensive biotechnology, vol. 1, Pergamon, Oxford, p 301
40. House WM, Shearer M, Maroudas NG (1972) Exptl. Cell Res. 71: 293
41. Spier RE (1985) In: Spier RE, Griffiths JB (eds) Animal cell biotechnology, vol. 1, Academic, London, p 243
42. Skoda R, Pakos W (1977) Belgian Patent 854,898
43. Carwell J, Burbridge C (1985) In: The world biotech report 1985, vol. 1 (Europe), On Line, London, p 83
44. Birch JR, Cartwright TB, Ford JA (1981) German Patent DE 3,031,617
45. Hurni WM, McAleer WJ, Hilleman MR (1978) British Patent 1,498,354
46. Girard HC, Suton M, Erdem H, Gurham I (1980) Biotech. Bioeng. 22: 477
47. Knazek RA, Gullino PM, Kohler PO, Dedrick RL (1972) Science 178: 65
48. Munder PG, Modonell M, Wallach DFH (1971) FEBS Lett. 5: 191
49. Jensen MD (1981) Biotech. Bioeng. 23: 2703
50. Lydersen BK, Pugh GG, Paris MS, Sharma BP, Noll LA (1985) Bio/Technol 3: 63
51. Earle WR, Schilling EL, Shannon JE (1951) J. Natl. Cancer Inst. 12: 179
52. Whiteside JP, Spier RE (1981) Biotech. Bioeng. 23: 551
53. van Wezel AL (1967) Nature 216: 64
54. Karkare SB, Phillips PG, Burke DH, Dean Jr, RC (1985) In: Feder J, Tolbert WR (eds) Large-scale mammalian cell culture, Academic, Orlando, p 127
55. Lim F, Sun AM (1980) Science 210: 908

56. Panina GF (1985) In: Spier RE, Griffiths JB (eds) Animal cell biotechnology, vol. 1, Academic, London, p 211
57. Oldshue JY (1983) Fluid mixing technology, McGraw-Hill, New York
58. Moore BE, Hansenpuch RE, Gerner RE, Burns AA (1968) Biotech. Bioeng 10: 625
59. Reuveny S, Zheng Z-B, Eppstein L (1986) Amer. Biotech. 4: 28
60. DeBruyne NA, Morgan BJ (1981) Internat. Lab. 11: 49
61. Techne Corporation, Inc., Princeton NJ (1985) The BR-05 Bioreactor, Catalog 20312
62. Läderach H, Widmer F, Einsele A (1978) In: Proc. First Europ. Congr. Biotech., part 1, p 84, Interlaken, Switzerland
63. Roubicek RV, Feres V (1987) U.S. Patent 4,657,677
64. Erickson LE, Stephanopoulos G (1987) In: Carberry JJ, Varma A (eds) Chemical reaction and reactor engineering, M. Dekker, New York, p 779
65. Brown PC, Costello MAC, Oakley R, Lewis JL (1985) In: Feder J, Tolbert WR (eds) Large-scale mammalian cell culture, Academic, Orlando, p 59
66. Spectrum Medical Industries, Inc., Los Angeles, CA (1987) Fleaker Hollow Fiber Concentration System, technical information
67. Tolbert WR, Lewis Jr C, White PJ, Feder J (1985) In: Feder J, Tolbert WR (eds) Large-scale mammalian cell culture, Academic, Orlando, p 97
68. Merten O-W, Reiter S, Himmler G, Schirer W, Katinger H (1985) Develop. Biol. Standard. 60: 219
69. Feder J, Tolbert WR (1983) Sci. American 248: 36
70. Germerdonk R, Ohlinger H-P (1987) Chem.-Ing.-Techn. 59: 244
71. Litz LM (1985) Chem. Eng. Progr. 81: 36
72. Spier RE, Whiteside JP (1984) Develop. Biol. Standard. 55: 151
73. Bajpai RK, Reuss M (1988) Chem. Eng. Commun., submitted
74. König B, Buchholz R, Lücke J, Schügerl K (1978) Ger. Chem. Eng. 1: 199
75. Bellco Glass, Inc., Vineland, N.J. (1985) The Bellco Bioreactor, technical information
76. Ku K, Kuo MJ, Delente J, Wildi BS, Feder J (1981) Biotech. Bioeng. 23: 79
77. Dostalek M, Haggstrom M (1982) Biotech. Bioeng. 24: 2077
78. Klement G, Scheirer W, Katinger HWD (1987) Develop. Biol. Standard. 66: 221
79. Bioreactor Technology, Inc., Schenectady, N.Y. (1987) MCBR Production System (distributed by C. Itoh Co.)
80. Feder J, Lewis C, Tolbert WR (1984) European Pat. 0113328
81. Mattiasson B (1983) In: Mattiasson B (ed) Immobilized cells and organelles, CRC Press, Boca Raton
82. Tyo MA, Hu W-S (1986) In: 192nd ACS Meeting, Division of Microbial and Biochemical Technology, Anaheim, CA, Sept. 7–12, 1986, paper 34
83. Tyo MA, Spier RE (1987) Enzyme Microb. Technol. 9: 514
84. van Wezel AL, van der Velden-de Groot CAM, de Haen HH, van den Henvel N, Schasfoort R (1985) Develop. Biol. Standard. 60: 229
85. Rebsamen E, Goldinger W, Schierer W, Merten O-W, Pálfi GE (1987) Develop. Biol. Standard. 66: 273
86. Zwietering TN (1958) Chem. Eng. Sci. 8: 244
87. Chudacek MW (1982) In: Stephens HS, Goodes DH (eds) 4th European Conf. on Mixing, Leeuwenhorst, The Netherlands, Cranfield, BHRA Fluid Engineering, p 275
88. Chudacek MW (1985) Chem. Eng. Sci. 40: 385
89. Chudacek MW (1985) Ind. Eng. Chem. Proc. Des. Develop. 24: 858
90. Buurman C, Resoort G, Plaschkes A (1986) Chem. Eng. Sci. 41: 2865
91. Bourne, JR, Sharma RN (1974) Chem. Eng. J. 8: 243
92. Musil L (1976) Collect. Czechoslov. Chem. Commun. 41: 839
93. Nienow AW (1985) In: Ulbrecht JJ, Patterson GK (eds) Mixing of liquids by mechanical agitation, Gordon and Breach, New York
94. Kay JM, Nedderman RM (1985) Fluid mechanics and transfer processes, Cambridge University Press, Cambridge
95. Muroyama K, Mitani Y, Yasunishi A (1987) Chem. Eng. Commun. 34: 87
96. Aeschbach S, Bourne JR (1972) Chem. Eng. J. 4: 234
97. Kneule F (1985) Internat. Chem. Eng. 25: 214

98. Chudacek MW (1986) Ind. Eng. Chem. Fundam. *25*: 391
99. Weisman J, Efferding LE (1960) AIChE J. *6*: 419
100. Einenkel W-D (1979) VDI-Forschungsheft *595*: 2
101. Rieger F, Novák V, Havelková D (1986) Chem. Eng. J. *33*: 143
102. Schulz R, Keaft H, Piehl GW, Lehmann J (1987) Develop. Biol. Standard *66*: 489
103. Adema E, Sinskey AJ (1987) Biotech. Progr. *3*: 74
104. Hallsby GA, Shuler ML (1986) Biotech. Bioeng. Symp. *17*: 741
105. Separation Equipment Technologies, Inc. (Setec), Livermore, CA. (1987) Tridentric (TM) Bioreactor, Cross-flow Products Bulletin, April 1
106. Pirt SJ (1975) Principles of microbe and cell cultivation, Blackwell, Oxford
107. Evans EA, Skalak R (1980) Mechanics and thermodynamics of biomembranes, CRC Press, Boca Raton
108. Einav S, Lee SL (1973) J. Multiphase Flow *1*: 73
109. Wichterle K, Kadlec M, Žák S, Mitschka P (1984) Chem. Eng. Commun. *26*: 25
109a. Schürch U, Einsele A, Kramer H, Widmer F, Eppenberger HM (1987) In: Neijssel OM, van der Meer RR, Luyben KChAM (eds) Proc. 4th European Congress on Biotechnology 1987, vol. 3, Elsevier, Amsterdam, p 569
110. Calderbank PH, Moo-Young MB (1959) Trans. Inst. Chem. Eng. *37*: 28
111. Reynolds AJ (1971) Thermofluid dynamics, Wiley, London
112. Nagata S (1975) Mixing: Principles and applications, Wiley, New York
113. Thomas DG (1964) AIChE J. *10*: 517
114. Matsuo T, Unno H (1981) J. Environ. Eng. Div., ASCE *107*: 527
115. Toni DT, Bagster DF (1978) Trans. Inst. Chem. Engrs. *56*: 1
116. Levins DM, Glastonbury JR (1972) Chem. Eng. Sci. *27*: 537
117. Batchelor GK (1960) The theory of homogeneous turbulence, Cambridge University Press, Cambridge
118. Shinnar R (1961) J. Fluid Mech. *10*: 259
119. Cherry RS, Papoutsakis ET (1986) Bioprocess Eng. *1*: 29
120. Friedlander SK (1957) AIChE J. *3*: 43
121. Rosenberg, MZ (1987) DSc Thesis, Washington University, St. Louis
122. Hinze JO (1971) Progr. Heat Mass Transfer *6*: 433
123. Das P, Kumar R, Ramkrishna D (1987) Chem. Eng. Sci. *42*: 213
124. Crougham MS, Hamel J-FP, Wang DIC (1987) Biotech. Bioeng. *29*: submitted
125. Kilburn DG, Webb FC (1968) Biotech. Bioeng. *10*: 801
126. Radlett PJ, Telling RC, Whiteside JP, Maskell MA (1972) Biotech. Bioeng. *14*: 437
127. Telling RC, Elsworth R (1965) Biotech. Bioeng. *7*: 417
128. Spier RE, Griffiths B (1984) Develop. Biol. Standard. *55*: 81
129. Aunins JG, Crougham MS, Wang DIC, Goldstein JM (1986) Biotech. Bioeng. Symp. *17*: 1
130. Mercer D (1981) Biotech. Bioeng. *23*: 2421
131. Clark NN, Atkinson CM, Flemmer RLC (1987) AIChE J. *33*: 515
132. Handa A, Emery AN, Spier RR (1987) Develop. Biol. Standard. *66*: 241
133. Donaldson TL, Boonstra EF, Hammond JM (1980) J. Coll. Interfac. Sci. *74*: 441
134. Frangos JA, McIntire LV, Eskin SG (1987) Biotech. Bioeng. submitted
135. Ben-Ze'ev A (1985) Exptl. Cell Res. *157*: 520
136. Napolitano EW, Pachter JS, Liem RKH (1987) J. Biol. Chem. *262*: 1493
137. Fleischaker RJ, Sinskey AJ (1981) Europ. J. Appl. Microbiol. Biotechnol. *12*: 193
138. McLimans WF, Blumenson LE, Tunnah KV (1968) Biotech. Bioeng. *10*: 741
139. Moser A (1973) Biotech. Bioeng. Symp. *4*(1): 399
140. Mukhopadhyay SN, Ghose TK (1978) J. Ferment. Technol. *56*: 558
141. Danckwerts PV (1970) Gas-liquid reactions, McGraw Hill, New York
142. Phillips KL, Sallans HR, Spencer JFT (1961) Ind. Eng. Chem. *53*: 749
143. Coppock PD, Meiklejohn GT (1951) Trans. Inst. Chem. Engrs. *29*: 75
144. Moser A (1977) Chem.-Ing.-Techn. *49*: 612
144a. Harakas NK, Lewis C, Bartram RD, Wildi BS, Feder J (1984) Advan. Exptl. Med. Biol. *172*: 119
145. Chen BH, Mallas K, McMillan AF (1985) AIChE J. *31*: 510

146. Hu W-S, Wang DIC (1986) In: Thilly WG (ed) Animal cell technology, Butterworths, Boston, p 167

147. Katinger H, Scheirer W (1985) In: Spier RE, Griffiths JB (eds) vol 1, Animal cell bio-technology, Academic, London, p 167

148. Lehmann J, Piehl GW, Schulz R (1987) Develop. Biol. Standard. *66*: 227

149. Phillips AW, Ball GD, Fantes KH, Finter NB, Johnson MD (1985) In: Feder J, Tolbert WR (eds) Large-scale mammalian cell culture, Academic, New York, p 87

150. Birch JR, Thompson PW, Lambert K, Boraston R (1985) In: Feder J, Tolbert WR (eds) Large-scale mammalian cell culture, Academic, New York, p 1

151. Banks GT (1979) In: Wiseman A (ed) Topics in enzyme and fermentation technology, vol. 3, Ellis Horwood, Chichester, p 170

152. Montagnon B, Vincent-Falquet JC, Fanget B (1984) Develop. Biol. Standard. *55*: 37

153. Nilsson K (1987) Trends Biotech. *5*: 73

154. Parenteral Drug Association, Inc. (1980) Technical Monograph No. 1, Philadelphia

155. 21 Code of Federal Register, 820.1–820.185, April 1, 1987

156. Harrison FG (1985) Bio/Technol. *3*: 43

157. Croy RRD, Fielding RM, Billiet CL (1984) Process Biochem. *19*(6): 209

158. Ball GD (1985) In: Spier RE, Griffiths JB (eds) Animal cell biotechnology, vol 2, Academic, London, p 87

159. Giorgio RJ, Wu JJ (1986) Trends Biotech. *4*: 60

160. Glacken MW, Fleischaker RJ, Sinskey AJ (1986) Biotech. Bioeng. *28*: 1376

161. Charles River Biotechnical Services, Inc., Wilmington, MA (1987) The Opticell Culture System, technical bulletin

162. Armiger WB, Forro JR, Lee J-F, MacMichael G, Mutharasan R (1987) On-line measurement of hybridoma growth by culture fluorescence, a communication, BioChem Technology, Inc., Malvern, PA

163. Fernandez EJ, Clark DS (1987) Enzyme Microb. Technol. *6*: 259

164. Drury DD, Dale BE (1987) Biotech. Bioeng. submitted

165. Aguayo JB, Blackband SJ, Schoenigo J, Mattingly MA, Hindermann M (1986) Nature *322*: 190

166. Sprague EA, Steinbach BL, Nerem RM, Schwartz CJ (1987) Circulation *76*: 648

167. Lambe CA, Walker AG (1987) In: Webb C, Mavituma F (eds) Plant and Animal Cells. Process Possibilities, E. Horwood, Chichester, p 116

168. Griffiths JB, Cameron DR, Looby D (1987) In: Webb C, Mavituma F (eds) Plant and Animal Cells. Process Possibilities, E. Horwood, Chichester, p 149

169. Fleischaker R (1987) In: Lydersen BK (ed) Large Sclae Cell Culture Technology, Hauser Publs., Munich, p 59

170. Macmillan JD, Velez D, Miller L, Reuveny S (1987) In: Lydersen BK (ed) Large Scale Cell Culture Technology, Hauser Publ., Münich, p 21

171. Lehmann J, Vorlop J, Buntermeyer H (1988) In: Spier RE, Griffiths JB (eds) Animal Cell

172. Wichterle K (1988) Chem. Eng. Sci. *43*: 467

173. Smith CG, Greenfield PF, Randerson DH (1987) In: Spier RE, Griffiths JB (eds) Modern Approaches to Animal Cell Technology (ESACT/OHOLO Conf.), Butterworth, London, p 316

174. Robertson B, Ulbrecht JJ (1987) In: Ho CS, Oldshue JY (eds) Biotechnology Processes. Scale-up and Mixing, American Institute of Chemical Engineers, New York, p 31

175. McQueen A, Meihac E, Bailey JE (1987) Biotechnol. Lett. *9*: 831

176. Hirschel MD, Gruenberg ML (1987) In: Lydersen BK (ed) Large Scale Culture Technology, Hauser Publ., Munich, p 113

177. Feder J (1987) In: Spier RE, Griffiths JB (eds) Modern Approaches to Animal Cell Technology (ESACT/OHOLO Conf.) Butterworth, London, p 454

Microbeads and Anchorage-Dependent Eukaryotic Cells: The Beginning of a New Era in Biotechnology

A. O. A. Miller[1], F. D. Menozzi[1], and D. Dubois[2]
[1] Laboratoire de Biochimie Moléculaire; [2] Unité de Biotechnologie Appliquée, Faculté de Médecine — Université de l'Etat 24, Avenue du Champ de Mars, 7000 Mons Belgique

Modern methods for the mass cultivation of anchorage-dependent mammalian cells started with the advent of microcarrier technology. Largely for reasons pertaining to their mode of preparation and ease of cultivation, 150–230 µm microbeads have been overwhelmingly adopted and the technology around them developed. To meet high biomass, macroporous microbeads have been developed. Also, the chemistry of the microsupport has been adapted in order to afford better protection of fragile cells to mechanical wear while simultaneously reorienting their differentiation towards the sought aims (production of cytokines, enzymes etc. . . .). Future progress depends upon solutions being brought to problems inherent to this new technology (maintenance of steady state conditions of growth etc. . . .) as well as to requirements arising from animal cell culture in general (biosensors, bioreactor's design etc. . . .). Besides such technical implementations, biology at large is also expected to benefit from the advent of microcarriers in fields as diverse as the preparation of metaphasic chromosomes in bulk, toxicity testing, organ reconstitution following cell transplantation etc. . . .

1 Introduction

The possibility of using the biosynthetic machinery of living organisms, to obtain molecules of therapeutic-, diagnostic- or industrial value, has considerably encouraged the development of biotechnology.

Extensive use of *E. coli* for such purposes has quickly led to the realization that the molecules thus produced were frequently not identical with the natural products.

Advances in Biochemical Engineering/
Biotechnology, Vol. 39
Managing Editor: A. Fiechter
© Springer-Verlag Berlin Heidelberg 1989

The differences resulted mainly from the absence of various co- and post-transla-tional modification mechanisms [36, 116, 127]. Due to differences in the abundance of specific tRNA and codon usage between higher and lower eukaryotes, yeast is not a perfect alternative [36]. There are even indications that glycoproteins can only be properly produced in animal cells in which they naturally occur [128]. For this reason animal cell culture in recent years has begun to attract much attention.

While it is generally accepted that animal cells constitute a good host, at least for the production of viruses and proteins of animal origin, the problem of producing them easily and cheaply on an industrial scale has deterred many from using them.

Mass cultivation of hybridomas, the first successful breakthrough, originated from the need to obtain large quantities of monoclonal antibodies (monoclonal antibody production is forecast to be a multimillion dollar industry). The technology adopted, i.e. growth in submerged cultures, did not fundamentally depart from the growth of bacteria in suspension. This made the transition from the bacteria to the animal cell culture easier as exemplified by the growth somewhat later, of alpha interferon-producing Namalwa cells of lymphoblastoid origin in an 8000 L reactor [97].

Not all mammalian cells, however, behave like hybridomas or Namalwas'. Many do not grow in suspension at all and require a support to attach to without which they will not grow.

"Anchorage-dependent" cells are cells from many transformed cell lines (CHO, VERO etc. . . .), tissue primary cells and cells from human diploid strains (WI-38 [46]; MRC-5 [55]; IMR-90 [88]; TIG-1 [91]). Notwithstanding the finite life span and high serum requirements of the two last categories of cells, two factors which severely limit their generalized use, their "normal" character and lack of oncogenicity made them the hosts of choice for the production of several vaccines.

Because of their immortality, aneuploid transformed cell lines such as BHK21, CHO, VERO and HeLa, present attractive alternatives. However, their oncogenicity measured by their capacity to induce tumours in nude mice or in newborn, X-irradiated syngeneic animals [147] have long barred them from being licensed for the production of biologicals for human use [94–96].

The detection of oncogenes in normal cells together with the development of new radioactive DNA probes able to detect contaminating DNA of lesser size than a potentially transforming oncogene [4] — have recently led to a progressive lifting [96, 148] of the ban imposed on the use of such established cell lines.

Growth of anchorage-dependent cells is usually realized in petri dishes, in Roux flasks or in large trays. Automation of the manipulations involved in the propagation of anchorage-dependent cells using roller bottles in series has made possible some scale-up in the production. This is labour- and time-consuming, therefore expensive, requiring highly trained and dedicated personel. Until some twenty years ago, the culture of anchorage-dependent cells was more an art than a science.

In his 1967 paper, where he showed that anchorage-dependent cells readily grow on the surface of microbeads kept in suspension by gentle agitation [133], van Wezel set the foundations of what was to become known as microcarrier technology.

Today [126], five hundred and even 1000 L microbead-based bioreactors are used [84]. This success stems from the fact that once anchorage-dependent cells have attached, the system can now be handled as a suspension culture in stirred tanks in much the same way as the hybridomas or the lymphoblastoid Namalwa cells mentioned above.

After a discussion of the several methodological aspects which condition the development of microcarrier technology, this paper will briefly review the biologicals already produced. Finally, the focus will be laid on new lines of investigation offered by this approach in order to anticipate new methodological breakthroughs which in turn could open new vistas in pure and applied research.

2 Present Status of Microcarrier Technology

2.1 General Outlines

With some adaptation due to the particulate nature of the microsupport, microcarrier technology lends itself to full environmental control of pH, pO_2, pCO_2. Control is made considerably easier, allowing good scale-up potential, while the user retains a considerable flexibility over the best surface-to-volume medium ratio to use for the optimized production in each specific case either of animal cells (or of biologicals extractable from them) or of secreted molecules.

As a rule of thumb, at confluence, one microbead (130 to 250 microns diameter), accomodates between 100 and 200 cells of 12 to 20 microns diameter. Depending upon cell size and on the commercial origin of the available microbeads, some 1.5×10^5 cells can be lodged on 1 cm^2 of microsupport[1], the number of microbeads per g of dry weight material varying between 2.5 and 5 millions.

Routinely, when the rules for proper inoculation are met, cultures inoculated with as few as 10^5 cells per mL of culture and 1 to 5 g microbeads per liter of final volume, reach cell yields of 10^6 cells per mL, i.e. one-thousandth of tissue concentration (1 mL wet packed cells per L of tissue i.e. 200 μg of cells) [5]. Results such as these can be obtained without much effort using conventional reactors run in batch mode. This, together with the potential for unlimited expansion explains the interest raised by this technology. In comparison, stationary cultures of actively growing anchorage-dependent cells (generation time around 20 to 24 h) have their yields increased 12- to 15-fold over a period of approximately 10 days [54]. This superior performance must be weighted against the volume of growth medium required (five times more cells are obtained per mL of microbead suspension as per mL of stationary culture), the limited growth area and the difficulties of scale-up.

Increasing the concentration of microbeads up to 10 to 30 g L^{-1}, a value approaching the maximum value theoretically attainable[2], allows cell yields around 10^7 cells per mL to be reached at the expense of greater complexity: such systems absolutely

---

[1] With N and D representing the number of microbeads per g of dry weight material and the mean diameter of the microsupport (in microns) respectively, the area of cultivation A in cm^2 per g can be calculated as of $A = 3.14 \times N \times D^2 \times 10^{-8}$.

Typically, 1 g of dry Cytodex 3 has an area of cultivation of:
$3.14 \times 2.65 \; 10^6 \times 200 \; 10^{-8} = 3328 \; cm^2$ i.e. 19–175 cm^2 plastic Roux flasks.

[2] One gram of completely swollen Cytodex 3 microbeads occupies 15 mL. The maximum concentrations of microbeads is thus $1000/1.5 = 66.6$ g L^{-1}, a value above which the microbeads will no longer be in suspension.

require perfusion, adequate oxygen supply [31] and careful optimization of the composition of the growth medium [9].

Despite these advantages offered by microcarriers it is important to stress that substances *secreted* by animal cells until now seem to be more readily obtained and in greater yields by immobilized cell cultures (hollow fibers [111, 146], microcapsules [90, 105], glass sphere propagators [144], tri-dimensional "honeybees ceramic networks [70], porous microcarriers [89] or static maintenance reactors [126]). This is not to say that microcarriers will not in the future be used for such a purpose but, their advantage is that they provide immediately accessible anchorage-dependent cells — and/or recovery of the cellular components which can be extracted from them.

The reduction of the lag phase and the increased growth extent observed with anchorage-dependent cells when the size of the inoculum is increased, has been repeatedly ascribed to metabolic cooperation among cells [49, 54, 104]. This results from the unhindered exchange directly from one cell into another, through closely apposed regions of the membrane (gap junctions [50, 103]) or by direct tactile signals [54], of small (800–1200 Da) molecules [138] and culminates in the case of transporting epithelia (renal tubule, intestinal mucosa) in the formation of a truly selective permeability barrier [98].

Since all populations experience a normal rate of mutation of 10^{-4} to 10^{-6} per allele per generation, anchorage-dependent cells have a greater chance to survive a mutation or a transient metabolic depletion than do cells propagated in suspension, the latter obtaining the missing metabolite from neighbouring cells only after their dilution in the environing growth medium.

Because cultivation of anchorage-dependent cells is still largely empirical, mathematical models have been developed to take these observations into account. It has been hypothesized [54] that cells in the inoculum attach to individual microbeads, adopting a Poisson distribution. They then strive for survival in the hostile environment which prevails under *in vitro* conditions of growth and succeed in doing so provided their number actually exceeds a certain critical value. Cells whose initial number is lower than this critical cell number exhibit retarded growth and reduced growth extent. Depending on the cell type, the kind of growth medium used and the chemical nature of the microsupport, knowing also the inoculum concentration, the microcarrier concentration and the microbead diameter distribution, it is now possible to calculate the distribution of cells on microbeads and from that the critical cell number.

2.2 The Microsupports

Although spherical microbeads have gained wide acceptance, very cheap positively laden cylindrical porous cellulose fibers can also be advantageously used [61, 68, 100, 102]. Interestingly, with this sort of support, while cells from established cell lines such as MDCK, Vero or BHK colonize the surface of individual fibres, human diploid fibroblasts do not: they grow, enmeshed within a 3-dimensional spheroid mass whose size increases in diameter as cell multiplication progresses [68].

Following changes of medium operated at regular times, the number of cells trapped in these spheroids increases 10-fold, before the culture begins to degenerate after 7 to 8, probably as a result of necrosis of the cells situated within the internal core of the aggregate.

The possibility of the cells, not only to be confined on the limited surface area outside the fibres which constitute the spheroids, but also to grow inside, allows growth with unprecedented high yields matched only by the recently developed macroporous microcarriers [89, 150]. This together with their low cost makes this type of microsupport very attractive indeed. The drawback presented by this support (as well as by macroporous microsupports in general) arises from the inaccessibility of the majority of the cells which are inside the microcarrier: such cells can become necrotic and evaluation of the biomass absolutely, depends upon prior trypsinisation.

Small disks 5 mm diameter and 150 µm thick, stamped from non-woven thermally bonded fabric sheets made from 10 to 12 µm diameter polyester fibres not only avoid the danger of necrotization just mentioned but also allow microcarrier cultures to be initiated with cell inocula containing as little as 5% of the final number of cells (as opposed to the 15 to 40% inoculum concentration required for initiation of a culture on microbeads) [16].

Because for a given volume, spheres offer the minimum surface area, microbeads have the worst geometry. However, the ease with which microbeads can be produced industrially as well as the dedicated hardware (spin filter, electronic sizing) which is being continuously developed around them as time passes, as a result, makes the introduction of microcarriers with a new geometry more difficult.

Most important, however, for the biology of the system is the chemical composition of the microbeads.

Depending upon the material used to make them, microbeads can be classified today into 6 categories: dextran, gelatin, polystyrene, polyacrylamide, glass and a sixth tentative one: composite microbeads. Their general characteristics are now given.

Dextran

Following the discovery by van Wezel [133] that DEAE-Sephadex A-50 could be used to propagate animal cells, the first modifications which were aimed to reducing its toxicity. This was achieved by lowering the ion-exchange capacity of this anionic exchanger from 3.5 meq g^{-1} down to 1 to 2 meq g^{-1} dry weight material (Cytodex 1 and 2, Superbeads), or by condensing the anionic charges within a pellicular rim limited to the periphery of the microsupport (Cytodex 2). Progressive characterization of the biological molecules intervening during anchorage (which can be broken down into four discrete yet overlapping steps which are adsorption — contact — attachment — spreading) led to the coating of existing dextran-based microbeads with collagen (Cytodex 3) and the manufacture of microsupports entirely made out of gelatin (Gelibead, Ventregel). Where very fragile cells are being cultivated, the possibility exists of avoiding unnecessarily proteolytic damage to the cells by dissolving dextran-based microbeads with dextranase [66]. Detailed description of the use of dextran microbeads have been published [77].

Gelatin

Microbeads made of plain gelatin have been developed to mimic in vivo conditions as much as possible and also because enhanced cell yields were obtained with some very fragile cells. The combined use of dispase and collagenase (together with EDTA

and low concentrations of trypsin) results in the preferential dissolution of the microbead matrix rather than in the loosening and destruction of the bonds which maintain
the cells attached. This leaves highly fragile cells unharmed. It is then sufficient to
pipet the suspension in order to obtain a homogeneous cell suspension ready to be
inoculated with new gelatin microbeads.

Serial propagation obviously now becomes a trivial problem [52]. The plating efficiency of cells propagated on such microbeads is considerably better than with cells
propagated on any other microsupport, because the trypsinisation step is omitted.

With the development of macroporous microbeads [89], it has been possible recently
to take further advantage of the superior biological properties of gelatin. Starting
from an emulsion in a hot (60 °C) gelatin solution of toluene droplets stabilized with
a water insoluble surfactant, astute manipulations yield spongy microbeads containing
crevices and macroporous cavities. The advantages offered by such a microsupport
are several: increased surface available for growth, protection of the cells against
mechanical stress, increased metabolic cooperation [138] and easy serial cultivation
following dispase-mediated microsupport solubilization. Along the same lines,
excellent results have also been obtained by Verax who developed a computer controlled fluidized bed reactor containing mammalian cells immobilized 5 cells deep, in
80% macroporous collagen microspheres 500 μm diameter. Since the same technology
can be applied to the manufacture of macroporous microbeads from other materials
such as polystyrene, polyacrylamide or dextran, it combines the advantages offered
by immobilized and suspension culture.

The limitations of these macroporous microbeads are the same as those of immobilized cell systems. There is a significant risk for those cells situated inside the beads of
becoming necrotic. Also, it is anticipated that steady state conditions of growth might
be exceedingly difficult to obtain.

Polystyrene

Because of the progressively increased sophistication of the microsupports and the
resulting increased costs incurred by their manufacture, alternative materials have
been investigated. Since most anchorage-dependent animal cell lines readily grow
in Petri dishes or in Roux flasks made of polystyrene, this hydrophilic material has
been an obvious choice for the manufacture of cheap microbeads. Careful comparative
studies made with underivatized polystyrene beads such as Biosilon or Cytosphere
have shown that the rate of attachment of diploid fibroblasts and of cells from established cell lines is low, a phenomenon probably reflecting the repulsion forces existing
between the microbeads and the cells which are both negatively charged.

Indeed, when positive groups (tertiary amines) are introduced during the polymerisation step depending upon the cell type tested, the cell yields increase some 9–27%.
However, despite these encouraging results, the yields are not better than those obtained with microbeads made of other materials [102].

The main drawback of polystyrene beads is their fast speed of sedimentation and
opacity: cells situated circumferentially can only be observed as a thin rim under the
light microscope.

In contrast to dextran or polyacrylamide-based microsupports microbeads made out
of polystyrene do not collapse during the dehydration procedure with organic solvents required for transmission or scanning electron microscopy. Properly derivatized,

they have been used to study the morphological changes associated with the attachment and spreading of HeLa- and endothelial cells [20, 125].

Polyacrylamide

The foundation of microcarrier technology laid down 20 years ago by van Wezel used a chromatographic support made out of dextran. Polyacrylamide-based microbeads were only developed some 10 years later. Maybe it is this historical lag which explains that the latter material never really caught up with the former and that today's experiments are conducted in the vast majority of cases with dextran-based microbeads. The physical chemistry of microbeads made out of polyacrylamide is as good as that of their dextran homolog: such microsupports can be coated with elongated hydrophobic arms [101], with collagen and glucosaminoglycans, while mechanical strength

Table 1. Types of microsupport available commercially

Manufacturer	Country of origin	Commercial designation	Chemical composition	Size (μm)	Cultivation area ($cm^2 \ g^{-1}$)
Microbeads					
Pharmacia	Sweden	Cytodex 1	DEAE-Dextran	130–220	3900–6000
Pharmacia	Sweden	Cytodex 3	DEAE-Dextran coated with collagen	100–190	3300–5100
Pfeifer 8 Langen	Germany	Dormacell	DEAE-Dextran	140–240	7000
Flow laboratories	USA	Superbeads	DEAE-Dextran	135–205	6100
Nunc	Denmark	Biosilon	Polystyrene	160–300	255
SoloHill Eng. Inc.	USA	Bioplas	Polystyrene	90–150 150–210	325 475
Lux	USA	Cytosphere	Polystyrene	160–230	unknown
IBF	France	Micarcel G	Polyacrylamide	160–250	unknown
Biorad	USA	Bio-Carriers	Polyacrylamide	120–180	4700
KC Biological	USA	Gelibead	Gelatin	115–230	3300–4300
Verax	USA	Verax	Collagen	500	a)
Ventrex	USA	Ventregel	Gelatin	200	unknown
SoloHill Eng. Inc.	USA	Bioglass	Glass	90–150 150–210	325 475
Galil	Israël	Acrobead (GEL, -COL, -PL, -DAH)	Gelatin, collagen, Poly-D-Lysine, Diaminohexane	150–250	500
SoloHill Eng. Inc.	USA	Collagen	Collagen-coated Polystyrene	90–150 150–210	325 475
Fibres					
Whatman	UK	DE-52/DE-53	DEAE-cellulose	80–400 μm long 40–50 μm wide	
Sigma	USA	QAE-cellulose	QAE-cellulose	6–800 μm long 10–35 μm wide	

a) These macroporous microspheres are claimed to reach packing densities of 2×10^8 and 4×10^8 cells per mL for hybridomas and fibroblasts respectively.

can be considerably reinforced following the methylation of one of the two carbon atoms present in the double bond of the N[Tris-(hydroxymethyl)methyl] acrylamide monomer molecule [17].

Polyacrylamide microbeads are slightly superior to dextran. For example, HeLa Oxford cells — normally refractory on dextran and mitotic HTC cells progress more rapidly on polyacrylamide microbeads during the M-Gl interphase [80]. Vascular endothelial cells also, kept most if not all their differentiation properties [106] on polyacrylamide.

Composite Microbeads

Using a terminology still open to modifications, we propose to put in this category those chemically heterogeneous spherical microbeads which simultaneously meet the two following criteria:

— made of a ground matrix obtained by polymerisation (dextran, polyacrylamide, collagen, polystyrene, polyacrolein, etc. . . .);
— contain additional biological molecules known to contribute to any one of the following steps: adsorption, contract, attachment, spreading and growth.

Collagen and charged groups of small molecular weight added so as to modulate the overall charge of the microbead being considered as standard constituents of the ground matrix, are excluded from this tentative definition, their presence not contributing to the "composite" character of the microsupport.

According to this proposed definition, none of the microsupports discussed earlier are composite. However, they become so once they are grafted with components of the extracellular matrix (laminins, fibronectins, glucosaminoglycans, etc. . . .) known to regulate adhesion, contact and attachment or with an oncogene as long as the latter intervenes in cell growth.

Microbeads whose nucleus is made of polyacrolein belong to this new and very promising category of composite carriers. Their ground matrix made of polyacrolein is obtained by the polymerisation of acrolein in the presence of a radical source and of an appropriate surfactant, conditions which favour the formation of reactive aldehyde groups in very high numbers.

Depending upon the particular manufacturer, these polyacrolein microspheres are then coated either with glass until the required density of $1.02–1.04 \, \text{g cm}^{-3}$ is reached (Bioglas) or with agarose (AG-Acrobeads from Galil Microbead) whose concentration can be made to vary between 2 and 5% [72].

The high binding capacity resulting from the presence of numerous reactive aldehyde groups, allow AG-Acrobeads to be further and easily derivatized with DEAE or free amino groups present in molecules such as gelatin, collagen, poly-D-lysine, diaminohexane, to obtain Acrobead-DEAE, -GEL, -COL, -PL or -DAH microcarriers respectively.

These tailor-made microsupports allow the growth of a wide range of animal cells with yields second only to the DEAE-derivatized cellulose. Much experimental evidence links differentiation, i.e. the extent to which certain specific proteins become timely expressed, to morphogenesis, i.e. the cell shape which depends upon the constituents forming the extracellular matrix (ECM).

To the best of our knowledge, we are, however, unaware of any systematic analysis performed in order to study the influence that curvature of the microbead and the existing cytoskeleton's bending might have on the metabolism. For vascular smooth muscle cells at least, attachment on 40 µm polystyrene beads covered with fibronectin does not impair either viability or the contractile response to angiotensin II [25].

A perusal of the data collected in Table 1 makes it obvious that the size — hence the curvature — of the commercial microbeads fluctuates within the 100 to 230 µm range. This situation results from a compromise between two conflicting requirements.

Using large microbeads obviously lengthens the time between inoculation and termination of the culture, reducing the number of batches necessary to obtain a given number of cells. Also, a reduced cell inoculum can be used [16].

On the other hand, because of the high specific area ($cm^2 g^{-1}$) characteristic of microbeads of small size and the corresponding increased area offered to cultivation, cell yields in this system are expected to be higher than those obtained with microbeads of large size.

Because the use of small microbeads shortens the time needed to reach confluence, this results in an increased frequency of manipulations.

The extent of manipulations involved by using microbeads of small size offsetting the anticipated increased cell yields explaining the adoption of microcarriers of intermediate (100 to 230 µm) sizes.

Having touched upon the problems raised by the size of the microbeads it is interesting at this stage to examine the reasons underlying the choice of a particular microsupport.

High manufacturing costs and the fact that microbeads in general can seldom — if ever — be reused[3], makes them a very expensive material indeed. So, except in cases where the industrial production of biologicals absolutely rests upon the mass cultivation of anchorage-dependent cells, suspension-adapted cells might be a more adequate material to start with [4]. Where growth of anchorage-dependent cells becomes necessary to obtain large numbers of cells or high amounts of product extracted from them, the choice of a given support depends very much on the cell performance due to anchoring.

For instance, cells from established lines such as the human nasopharyngeal KB carcinoma, form multiple layers on the surface of glass microbeads but fail to do so on DEAE-dextran microcarriers (Cytodex 1) [135].

MRC-5 cells placed in presence of anionic exchange cellulosic fibres fail to colonize the surface of the microsupport under stationary conditions [102] but form spheroids in suspension and embed in the fibers [4, 68].

Finally, in stationary cultures of derivatized cellulose fibres, cells from many established lines (BHK, MDCK) grow as monolayers as opposed to primary cells such as chick embryo fibroblasts or rat embryonic neuronal or muscular cells which form multilayers [102].

These three examples show that not only the chemical composition of the microcarrier, but also mode of culture (stationary-versus agitated-cultures) and ploidy of the cells influence growth.

---

[3] Bioglas can be used twenty times or more.

Comparative scanning electron micrographic studies made on KB [135] and human diploid or first passage monkey kidney cells [136] propagated on glass or on Cytodex 1 microbeads agitated in submerged cultures show consistent morphological differences taking place irrespective of the cell type but depending only upon the kind of micro-carrier used. On glass, the cells are considerably flattened, their surface is studded with many microvilli and their anchorage relies on the presence of long, slender filopodia.

On soft Cytodex 1 microbeads the filopodia completely disappear. The cells are much thicker and they adhere so strongly as to deform the entire edge of the cell making it appear slightly imbedded in the soft deformable texture of the matrix.

Such differences in morphology must reflect differences in the strength with which cells adhere to their support. Indeed, cells grown on glass can be much more easily detached using a milder trypsinization than cells propagated on Cytodex 1 which requires a more drastic treatment and therefore display a lower efficiency of plating.

Also, a comparison between the levels of proteolytic enzymes secreted by cells propagated on glass microbeads or on Cytodex 1, showed the latter medium to contain higher amounts [136, 137].

Leukotrienes, produced by the action of 5-lipoxygenase on arachidonic acid, confer glass non-adherence on leukocytes from either cancer-bearing patients [125] or asthmatic individuals [30]. This characteristic is probably due to interference with the roles played by certain glycoproteins [3]. In line with these observations, MRC-5 human lung di-ploid fibroblasts grown on glass microbeads show the highest levels of leukotrienes B4 and C4.

These results confirm earlier experiments that changing the chemical composition of microbeads drastically affects cell shape (morphology) [142] which in turn profoundly influences cytoplasmic organization, cell metabolism and the overall program of cell differentiation [12]. The implications of this open for the microbeads new and fascinating fields of application that other modes of cultivation do not.

Depending upon the cell type, factors in the growth medium also alter such micro-support-cell interactions and superimpose their effects on those just evoked. Together with serum-free chemically defined medium, microbeads already contributed to a better understanding of the factors underlying cell differentiation and morphogenesis. A detailed analysis of the effects induced by manipulating the chemical composition of the growth medium is however outside the scope of this paper.

2.3 Technical Implementations

2.3.1 In Situ Evaluation of Biomass

In submerged cultures of exponentially multiplying cells kept under steady state conditions of growth, the amounts of glucose and lactate formed at any given time are proportional to the cell concentration. Provided such growth conditions are rigor-ously maintained and the passage from one cell density to another is made slow enough so as to allow the ensuing fluctuations to be truly reversible in a thermo-dynamic sense, then the appropriate glucose and lactate probes can be used in order to measure biomass. Besides the difficulty sometimes of strictly adhering to the above conditions, it must be realized that any modification brought upon such a system (such as the amount of oxygen available to the cells) bears the risk of the glucose/

lactate concentrations being no longer proportional to the cell density. Besides these limitations, the fact that construction of enzyme sterilizable electrodes ("biosensors") for in-situ monitoring is fraught with many difficulties, makes this type of measurement very problematic.

The same metabolic limitations apply of course in the case where evaluation of the biomass is based on a measure of the fluorescence intensity emitted upon illumination at 340 nm of intracellularly-located reduced nicotinamide adenine diphosphonucleotide NAD(P)H. The fact that phenol red (present as an indicator in most culture media formulations) and riboflavin also fluoresce at the same wavelength complicates the situation even more although the possibility offered by these probes of tuning to other wavelengths and of being autoclavable, represents considerable progress.

The evaluation of the cell concentration in submerged cultures is difficult, especially in situ[4] in microbead-based bioreactors. It is almost impossible to deal with this problem, and the old off-line procedure of Sanford [107] as modified by van Wezel [134] is still the most adequate method.

The situation as it stands today is just the opposite.

Using standard equipment such as the Coulter Counter, a 560-microns orifice glass probe and the multichannel pulse height analyzer recently developed by the same company (C256 Channelyzer), it is possible, after appropriate dilutions of the growth medium, to obtain the size distribution histogram of cell-laden microbeads. The time-dependent shift of the individual distributions as measured by their increased mode (Peak Channel Number: PCN) observed during growth, varies in proportion with the increased number of cells present on individual microbeads[5]. Not only is it possible in this way to estimate in-situ the cell density of exponentially growing cultures [81, 122] but, when cells become confluent, the changes in the way they dispose themselves on the microbeads (flat multipoint attachment versus stalk-like adhesion) and also cytopathic effects can be readily detected [44].

Further miniaturisation of this probe will make it implantable. Implemented by dedicated hardware with specific software being developed for each of the cell types most commonly used today (VERO, MDCK, CHO, HeLa, . . .) it will transform cell enumeration, a hitherto complex procedure, into a trivial, fully automated measurement.

While the angular coefficients of the straight lines obtained by plotting cell density versus PCN vary with the absolute size of the cells attached [122] (as measured under the microscope) it is by no means certain that variations in PCN actually reflect fluctuations in size only. Although "electronic sizing" is the most common and convenient denomination of this type of measurement, recent results seem to suggest that progressive increases in electric impedance offered to the passage of the current through the 560-microns orifice by the cell-laden microbeads, results from a progressive increase in electric resistance offered by the cells which constitute as many "ab-

---

[4] The term in situ is used to describe the situation where measurements are made on cells while these remain attached to the microsupport. Direct sensing in the reaction mixture is identical with on-line measurement.

[5] The average number of cells present per microbead is equivalent to ϱ, the cell density (cells per microbead). Cell concentration refers to the number of cells present per unit volume of culture medium (cells per mL).

solute resistors". If this hypothesis is correct, it will allow us to follow in-situ changes affecting the membrane shape and electric permeability (binding of growth factors on their specific receptor, action of certain antibiotics such as coumermycin, etc. . . .). Possible applications of the use of this probe in toxicology coupled to the microbead technology are given in Sect. 3.2.

2.3.2 Influence of Bioreactor's Design on Shear Stress

When cell lines have been genetically optimized, bioreactor design becomes determinant [51]. Among the critical factors affecting cell growth, shear stress has been acknowledged by everybody working with microbeads as being of the utmost importance.

The first and still most widely used approach in this field is an empirical one [49, 53]. Recording the growth rates of cells propagated on microbeads over a wide range of agitations, allows to express the growth rate as a function of the Integrated Shear Factor (ISF) [53].

Unexpected results, such as the finding that certain combinations of impeller size and vessel diameter make cell growth independent of impeller tip speed [53], emphasize the potentialities of this type of approach.

As useful as it is, the ISF which measures the shear field between the propeller and the vessel's inner walls is no reliable basis for the scale up of bioreactors with similar or dissimilar geometry [22].

A first step towards a global mathematical model is based on the fundamental equations of fluid mechanics and rests on the assumption that the detrimental effects of shear on growth result from random interactions of cells with eddies [22].

The first results obtained by this mathematical approach provide the manufacturer of bioreactors with some engineering guidelines for better design with considerable savings.

As the need to produce more cell biologicals at lower prices increases, new mathematical approaches will inevitably be developed (e.g. statistical thermodynamics) which will extend and synergistically complement those developed earlier.

2.4 Current Production of Biologicals by Cells Propagated on Microbeads

Anchorage-dependent cells propagated on microbeads have been repeatedly used at the laboratory scale (100 ml to 5 L) for the preparation of viral antigens [29, 47, 145] and for the isolation of molecules secreted in the culture medium such as interferon β from mouse L cells [23], angiogenesis factors [33] and plasminogen activator from human foreskin cells etc. . . .

Because of the several technical difficulties encountered during the scaling up, comparatively few such processes have yet been transposed on an industrial scale. Table 2 gives a list of some of them.

In the coming years, the number and diversity of biologicals produced on an industrial scale by either "normal" or genetically engineered mammalian anchorage-dependent cells will increase considerably [99]. This number however constitutes only a small fraction of all biologicals amenable to industrial production using microbeads [21, 41, 83].

Table 2. List of some biologicals obtained on an industrial scale from anchorage-dependent cells propagated on microbeads

Biological	Type of cell used	Volume of culture (L)
Tissue plasminogen activator	Guinea pig keratocytes	17.5 [40]
Human β interferon	Genetically modified mouse L929 TK cells	22 [112]
Erythropoeitin	Genetically modified CHO cells	unknown [129]
Aujeszky vaccine	Swine testicular NLST established cell line	150 [7]
Foot-and-mouth vaccine	Pig kidney cell line IBRS2	150 [74]
Polyomyelitis vaccine	VERO	1000 [85]

3 Future Trends

The foregoing examples made it clear that microbeads have been used until now essentially as a mere mechanical substitute affording a convenient way of concentrating cells. Utilization of microbeads for such purposes is so obvious that this trend will undoubtedly continue to develop during the forthcoming years. Limiting ourselves to these aspects only, would leave the potentialities of microbeads largely untapped.

Indeed, ongoing experiments in basic research, rely more and more on the use of microbeads to study cell-cell and cell-support interactions. It is expected that the knowledge acquired along these lines will be quickly channeled to more applied research: adaptation of the biochemistry of the microbead and that of the growth medium will then allow the biotechnologist to harness at will the cell's metabolic potential to his own aims.

The following examples illustrate some of the developments expected to happen in the field of microbead-based technology.

3.1 Metaphasic Chromosomes

Using recombinant vectors and one of the several methods developed to introduce them in suitable recipient hosts (transfection, DNA-mediated gene transfer, injection, use of transgenic animals [37], spheroplast fusion [19], injection [60], etc. . . .), it has been possible to obtain genetically manipulated cells stably expressing a given phenotype. This well-established technique, extensively applied in biotechnology, has allowed the production of several biologicals of interest.

Besides the low number of cells, expressing the new phenotype, which can be obtained by these methods, a situation which can possibly be remedied by cloning and isolation of the high producers only, the genes thus transferred are expressed poorly if at all and do not respond correctly to normal stimuli, a situation ascribable to the impossibility of appropriately targetting the genes to their specific sites in the recipient host [130].

Thus, at present, this approach is not directly applicable to the cure of human diseases by gene therapy.

Several examples indicate however that properly regulated gene expression can be observed when intact chromosomes are transferred such as during hybridization of somatic cells for instance. Also clear gene dosage effects can be observed with intact surnumerary chromosomes such as in the case for cytoplasmic superoxide dismutase gene present on chromosome 21 in Downs syndrome — afflicted individuals [43]. Wilm's tumour [143], retinoblastoma [63], and Bilateral Acoustic Neurofibromatosis [114] (BANF), are tumours associated with the specific loss of both of certain alleles situated on chromosomes 11, 13 and 22 respectively, an observation confirming earlier results that certain chromosomes contain genes whose presence is required in order to suppress malignancy [45, 57]. In line with these observations, recent experiments where Wilm's tumour cells were microcell-injected with chromosome 11, showed suppression of malignancy [143].

It is clear from these examples that disposing of pure, sorted chromosomes of each type in sufficient amounts (provided means exist to inject them into cells without breakage [79]), would constitute a formidable asset in genetic therapy.

Going one step further, it is possible to construct synthetic mammalian chromo-somes using the same techniques as for artificial yeast chromosomes [15, 86]. Man-made chromosomes containing DNA essential for chromosome segregation (centromere), replication (autonomously replicating sequences), stabilization (telomeres) and with amplifying target products genes could be obtained.

Such an approach clearly will revolutionize biotechnology.

Techniques to sort metaphasic chromosomes at rates in excess of $20.000 \ s^{-1}$ [93] and to separate chromosome-sized DNA molecules by pulse field gel electrophoresis [62, 132] are in routine use, while DNA-sequencing supercentres with a processing output of one-million bases a day are progressively being set up [64, 65].

The problem here is that today we cannot obtain the chromosomes in bulk quantities needed for such a molecular dissection of the human genome and that this, in fact, conditions the whole project.

Anchorage-dependent cells upon reaching mitosis become rounded and can be easily dislodged from their monolayer following brisk agitation of either the Petri dish or the Roux flask used for their propagation[6]. This is the well-known mitotic shake off procedure developed some 25 years ago by Terasima and Tolmach [121].

At any one given time, only 2–4% of the cells of an asynchromous population are in mitosis. This, with the limited surface available, makes the number of mitotics obtainable by this procedure pathetically small, therefore drastically limiting the yields of metaphasic chromosomes.

Mitotic cells can be easily dislodged from microbeads [82, 87] and their yields consid-erably increased by scaling up, using prior disruption of microtubules and accumula-tion of cells in mitosis [151] or following careful manipulation of the cell cycle [80].

Production of chromosomes in bulk, not only from human cells but also from other cells of different origins (insects, fish, ovine, bovine, avian, etc. . . .), will constitute the major development of medicine and biotechnology in the forthcoming years.

Obtaining chromosomes and other intracellular components is not the only possi-

---

[6] Complete detachment of the cells from their microsupport and their serial propagation on micro-beads [36], one of the major difficulties of microcarrier technology can be obtained — depending upon the cell type used —, by vortexing [106], avoiding the use of proteolytic enzymes [14, 67] and the sequelae they may induce on fragile cells.

bility offered to the biotechnologist by anchorage-dependent cells propagated on microbeads. Although still speculative for the time being, another very exciting utilization is the obtaining of all the organelles and biologicals required for preparative cell free protein synthesis (ribosomes, messenger RNA etc. . . .) [103a].

3.2 Culture of Cells on Microbeads in Toxicity Evaluation

Some 20 million laboratory animals are sacrificed each year in the USA alone [2, 113] with proportionately equivalent numbers in other industrialized countries of the world. This is in order to gain better toxicological information concerning only a hundred chemicals or so in current use.

Since the total number of chemicals in current daily use tallies around fifty thousand with a yearly increase of 1,500 to 2,000, full toxicity evaluation clearly becomes a daunting if not an impossible task to achieve by today standards.

The growing ethical concern of society together with the significant methodological advances achieved by in vitro cultures and the fact that the methods using living animals developed 50 years ago are very expensive and cumbersome, not to say anything about the scientific values of the results thus obtained, have led to a multiplication of the number of attempts aiming at validating in vitro procedures using animal cell cultures [38].

For this approach to succeed, a suitable quantifiable cell "descriptor" ought to be found for which an endpoint is clearly defined and whose variations parallel the toxic effects hitherto observed with the animal. Besides, the method should be rapid to perform and economically competitive.

The Draize eye irritation test is a good example of the alternative role in vitro cell cultures may play as well as of the contribution of microbeads to this particular field of toxicology.

The test consists in placing 100 µL of the chemical solution whose toxicity is to be evaluated, in one eye of an albino rabbit, leaving the other intact as a control, the degree of toxicity being assessed by the extent of eye irritation such as redness, swelling lacrymal secretions, etc. . . . Such measures are subjective, the results obtained being unreliable, difficult to compare one with another, thus making the pain inflicted upon the animal seem the more cruel. Quantitative uridine incorporation studies made with mouse embryonic cells cultivated in vitro [118] and its latest adaptation [26] based on the simultaneous use of HepG2 cells and [^{14}C] thymidine to measure cell quantity, seem to correlate well with the Draize eye irritation test.

A further improvement consists in finding alternative cell descriptors which do not rely on the use of radioactive isotopes.

It has already been shown that electronic sizing of microbeads covered with anchorage-dependent cells detects subtle cytopathic effects induced by the intracellular multiplication of the sexually transmissible parasite *Chlamydia trachomatis* [44]. Extensive comparative studies will tell us if variations of the overall size of the microbeads (which changes with the cube of the diameter) as deduced from PCN measurements, can match the sensitivity and reproducibility of the radioactive method and the results thus obtained be extrapolated to man.

In many area of toxicology such as evaluation of acute toxicity, skin irritancy tests, detection of tumour-promoting activity etc. . . ., microbeads offer unique possibilities of investigating the behaviour of vascular and pulmonary endothelial cells during their homeostatic regulation of the levels of blood angiotensin II, bradykinin, etc. ... [106]. Such systems can play a pivotal role in the elucidation of the effects polluted air (car exhaust fumes, tobacco smoke) has on the lungs.

3.3 Cultivation of Differentiated Cells for Transplantation

Because of the steadily increasing shortfall between the number of organs available for transplantation and the actual demand, there is a growing and urgent need to find alternative solutions.

Liver and skin offer a good example of the evolution towards this goal.

Repeated attempts at restoring crippled hepatic functions by injecting suspensions of isolated liver cells have met with limited success. Only when liver cells are microencapsulated [149] or when purified hepatocytes are first made to attach to microbeads [24], can hepatic metabolic functions be transiently restored [1].

The first advantage of using microbeads, is that cells being on the outside, allow the homing process to take place. The second advantage is as follows: liver failure may result from an impairment of metabolic functions normally fulfilled by only one of the several cell types which together contribute to the organ's physiological role in the body.

For instance, in certain strains of rats, the effects resulting from the absence of both alleles of the gene coding for bilirubinglucuronide transferase (increased levels of bilirubin in the plasma), can be transiently corrected — and a decrease in the levels of plasmatic bilirubin observed — following the injection of hepatocytes enriched laden microbeads [24]. Where the fat-storing, stellate, non-parenchymal cells[7] found in minor amounts in the liver and responsible, among other actions, for the elaboration of the hepatic matrix, become, for one reason or another, incapable of fulfilling their task, it is expected that injection of lipocyte-laden microbeads [34] would restore the liver matrix, a function they clearly would have been unable to perform if microencapsulated.

A significant contribution can be made by microbeads in the case of human liver transplantation [11] as indicated by the experiments with animals. There is now a potential possibility to avoid heavy surgery with its inherent costs and risks[8].

As shown by the recent successful implantation in mice of genetically manipulated fibroblasts [117] and of placental cells in patients afflicted by the Niemann-Pick disease [108], transplantation is not limited to one cell- nor to one type of genome.

This, together with the possibility of relying on a bank of cells whose histocompatibility groups have been determined well in advance, in which it would suffice to choose the best matching ones, opens fascinating new possibilities.

Gradual improvement of culture procedures especially the development of a culture medium containing very little if no serum at all [142 a], has extended the life expectancy of mouse embryo cells beyond the moment when such cells, in presence of serum,

---

[7] These cells are also called Ito cells or lipocytes.
[8] Some 2–4% patients fail to support organ transplantation and unexpectedly die during the operation.

usually undergo crisis. Although murine cells distinguish themselves in many respects from human ones, if this observation is confirmed, it would allow the stockpiling in a liquid nitrogen repository, directly on microbeads, of sufficient numbers of cells of each histocompatibility group, ready for microbead-mediated mass transplantation[9].

The high frequency of trauma affecting skin (chronic ulcers, wounds, burns, etc. . . . .) together with the inevitable rejection which accompanies allo- or xenografts, makes this organ to one of the most essential [27].

As will be now shown, it is also the organ whose transplantation will, in all probability, be the first to benefit from microbead technology.

Skin rejection results mainly from the host's immunological surveillance mechanisms of HLA-DR antigens present on the surface of transplanted Langerhans cells of the dermis. Except in pathological cases [6, 39, 59] or following treatment with gamma interferon [10, 71, 109, 140], keratinocytes which form the majority of the epidermal cells, do not express this antigen. The few remaining cells which do so are melanocytes and Langerhans cells but these do not survive conditions aimed at propagating keratinocytes in vitro. Grafting across a major histocompatibility barrier, keratinocyte-epithelia obtained in this way instead of whole skin biopsies, explains their success, when used as allograft in immunosuppressed adult patients, to cure extensive burns or cover large wounds [28, 123, 124].

Scaling up the culture of keratinocytes is hampered by the necessity of manipulating large numbers of plastic dishes aseptically.

Typically [123] the 4 to 6 millions or so keratinocytes present in a human biopsy of 2 cm^2 are used to inoculate 4 to 6 75 cm^2 Petri dishes containing a feeder layer made of 1.5 million X-irradiated 3T3 Balb/c murine fibroblasts. After 3 to 4 weeks, following several subcultures, this number of cells, amplified some ten thousand-fold, covers an area of 1 m^2 approximately.

Preliminary results show that human keratinocytes attach lightly and then multiply onto collagen-coated Cytodex 3 microbeads distributed in a ninety-six well tissue culture plate so as to uniformly cover the bottom of the individual wells [58].

These rather rudimentary experimental methods are applicable for mass scale cultivation of keratinocytes in keratinocyte-adapted [18], serum-free [8, 11, 13, 69], chemically-defined media, under steady state conditions of growth, monitored by a biomass electronic sizing probe.

Whether used for biotechnological purposes (obtention of biologicals of interest) or for transplantation, microbeads can have their surface biochemistry modified in such a way as to influence and possibly redirect the differentiation program of the cells which are grown upon [73, 139]. Several experiments lend support to the hypothesis that fibronectin known to regulate the degree of cell spreading[10], controls DNA

---

[9] Interestingly enough, cells (mammalian, insects) immersed in liquid nitrogen resist the freeze — thawing procedure better when they are attached on microbeads [78], allowing for the direct injection into the organism of the thawed cells without the need of an intermediate trypsinization step. Also, there are hints that the extent and specificity of the immunologic response are better when animals are injected with cells attached on microbeads than with isolated cells obtained by trypsinisation.

[10] Fully spread cells progressively cease with DNA synthesis after having attained a spherical form [32].

replication, protein synthesis being influenced by the presence of laminin which sets the extent to which cells establish contacts with- and attach to- the substratum [12].

Adding their action on top of that produced by fibronectin and laminin, constituents of the extracellular matrix also contribute to the final shaping of the cell and to the ensuing cytoskeletal reorganization.

For instance, vascular endothelial cells cultivated on type IV collagen become organized into highly differentiated tubelike-one of the manifestations of angiogenesis [46] — while on collagen type I or III, the same cells actively proliferate to form a confluent monolayer and only rarely show the microvascularity observed above [46]. Obviously the outcome of the transplantation of such cells depends very much on the type of collagen used.

3.4 On-line Sensors for Bioprocesses

Although it is highly probable that adequate probes will ultimately be constructed [48, 75, 76, 131, 141] enabling the biotechnologist to specifically measure the changes of concentration of whatever molecule he desires, miniaturization of the hundred or more sensors needed for a global evaluation of the bioprocess forces us to ask whether alternative, cheaper routes, more immediately available to us do not exist.

In fact they do.

One promising approach — others exist — relies on spectroscopic analysis in the 400 to 4000 cm^{-1} — range of wavelengths of the culture medium followed by adequate mathematical treatment of the information thus collected, the so-called Fourier Transform InfraRed spectroscopy (FTIR) [42].

The infrared part of the spectrum contains the information pertaining to the vibrational and rotational motions of amino-, amido-, carboxyl- and keto-groups present in almost all biological molecules.

Until recently high water spectral absorption (especially in the 1620 to 1690 cm^{-1}) and the very low energy released following the interaction of these groups with their environment, led to very low signal to noise ratios and made detection of particular spectral lines particularly difficult. However, spectral substraction and multiple scanning, together with the development of optic cells made out of insoluble and more transparent materials, have overcome these difficulties [119, 120].

Spectral resolution which, depending upon the spectrometer being used, can reach 0.002 cm^{-1}, allows the assignment of bands to the presence of certain chemical groups in β-pleated regions, α-helices or other regions of ill-defined structure [92]. More often, resolution is in the range of 8 cm^{-1} allowing as many as 25 spectra per s to be recorded and separately stored [42]. Such speeds allow subtle time-dependent structural modifications to be detected. Also, the progressive degradation of molecules taking place within the reactor (hybridomas secrete proteases which may act upon the monoclonal antibody just produced [110]) as well as enzymatic-affinity reactions taking place on suitably treated surfaces are within reach.

Until now, individual components present in mixtures had first to be fractionated (either by HPLC or gas chromatography) before being analyzed by FTIR. With increasing speed of data processing and spectral resolution characteristic of modern instruments, coupled to gated substracting procedures, it is possible to envisage the implantation of sterilizable on-line FTIR probes able to monitor simultaneously, qualitatively and quantitatively the composition of the growth medium used to cultivate animal cells.

4 Conclusions

Microcarriers, by allowing the large-scale cultivation of anchorage-dependent animal cells, significantly contributed to the industrialization of vaccine production. Also they made possible the use of diploid cells with finite lifespan — the only ones to be admitted by the regulatory authorities at that time — of molecules otherwise impossible to obtain.

Because of the ease with which spherical microcarriers can be manufactured, and their various uses, microbeads have gained wide acceptance among scientists and biotechnologists.

Anchored on microbeads, animal cells differ from immobilized ones in several important respects.

Firstly, the physical support provided by the microbeads protects the fragile cells against shearing forces (less though than for immobilized cells). Simultaneously, the surface biochemistry of the microsupport (together with the chemical composition of the growth medium) can be modified in such a way as to reorient differentiation of the cells and put their behaviour more in line with the sought aims [32].

Secondly, formation of numerous gap junctions between neighbouring cells in close apposition and the resulting exchange of metabolites further stabilize the cells by metabolic cooperation. Clearly, microbeads lend themselves ideally to the study of the interactions which take place between cells of different evolutionary origins as happens for instance during the formation of specialized tissues (parenchymeepithelium interactions) whether these cells are on the same microbead [115], in direct contact or separated from each other on different microbeads and then reunited.

Thirdly, at all times, the cells remain accessible to the experimenter, allowing not only the recovery of secreted molecules but also the isolation of intracellular components of interest and the availability of the required number of cells necessary to perform transplantation. The possibility of injecting differentiated or genetically manipulated cells still attached onto their (biodegradable) microsupport in order to cure physiological disturbances opens new and fascinating perspectives in medicine.

For all these reasons, it is evident that microbeads and anchroage-dependent eukaryotic cells open a new era in biotechnology.

5 Acknowledgements

We are much indebted to Dr. T. Clark for revision of the English.

6 References

1. Agius L, Battersby C, Alberti GMM (1985) In vitro Cell and Develop. Biol. 21: 254
2. Alternatives to animal use in research, testing and education. Congress of the United States. Office of Technological Assessment. Washington DC 20510, 1986

 3. Anderson DC, Springer TA (1987) Ann. Rev. Med. 38: 175
 4. Arathoon WR, Birch JR (1986) Science 232: 1390
 5. Arathoon WR, Telling RC (1982) Develop. biol. Standard 50: 145
 6. Auböck J, Romani N, Grubauer G, Fritsch P (1986) Brit. J. Dermatol. 114: 465
 7. Baijot B, Duchene M, Stephenne J (1985) Develop. biol. Standard 66: 523
 8. Barnes D, Sato G (1980) Cell 22: 649
 9. Barngrover D, Thomas J, Thilly WG (1985) J. Cell Sci. 78: 173
10. Basham TY, Nickoloff BJ, Merigan TC, Morhenn VB (1984) J. Invest. Dermatol. 83: 88
11. Benhamou J-P (1987) Presse médicale 16: 705
12. Ben-Zéev A (1986) TIBS 11: 478
13. Bertolero F, Kaighn ME, Camalier RF, Saffiotti U (1986) In vitro Cell Develop. Biol. 22: 423
14. Billig D, Clark JM, Ewell AJ, Carter CM, Gebb C (1984) Develop. biol. Standard 55: 67
15. Blackburn EH (1985) TIG. Jan. 8–12 p 8
16. Bohak Z, Kadouri A, Sussman MV, Feldman AF (1987) Biopolymers 26: S205
17. Boschetti E (1987) Personal communication
18. Boyce ST, Ham RG (1985) J. Tissue Cult. Meth. 9: 83
19. Brzeski H, Chambers M, MacDonald C, Stimson WH (1985) Develop. biol. Standard 60: 105
20. Burke D, Brown MJ, Jacobson BS (1983) Tiss. & Cell 15: 181
21. Collins J, Betz H, Davies JE, Fiers W, Paoletti E, Pfaff E, Rowlands DJ, Schallen H, Siewert G,
    Sippel AE, Wagner EF (1985) Animals and medicines. In: Silver S (ed) Biotechnology: poten-
    tials and limitations. Dahlem Konferenzen — Life Science Research Report 35 Springer-Verlag
    Berlin Heidelberg New York, p 127
22. Croughan MS, Hamel JF, Wang DIC (1987) Biotech. Bioeng. 29: 130
23. Delzer J, Hauser H, Lehmann J (1985) Develop. biol. Standard 60: 413
24. Demetriou AA, Whiting JF, Feldman D, Levenson SM, Chowdhury NR, Moscioni AD, Kram M,
    Chowdhury JR (1986) Science 233: 1190
25. Dennis PA, Wolley R, Taylor NS, Moyer CF (1986) Cytometry 7: 384
26. Dierickx PJ (1987) Arch. Intern. Physiol. Bioch. 95: B138
27. Dodet B (1986) Biofutur 52: 35
28. Faure M, Mauduit G, Schmitt D, Kanitakis J, Demidem A, Thivolet J (1987) Brit. J. Dermatol.
    116: 161
29. Fiorentine D, Shahar A, Mizrahi A (1985) Develop. biol. Standard 60: 421
30. Fink A, Bibi H, Eliraz A, Tabachnik E, Bentwich Z (1985) Immunol. Lett. 10: 319
31. Fleischaker RJ Jr, Sinskey AJ (1981) Europ. Appl. Microbiol. Biotechnol. 12: 193
32. Folkman J, Moscona A (1978) Nature 273: 345
33. Folkman J, Klagsbrun M (1987) Science 235: 442
34. Friedman SL, Roll FJ (1987) Anal. Biochem. 161: 207
35. Furcht LT (1986) Lab. Invest. 55: 505
36. Gannon F (1985) The choice of host organisms for industrial genetic engineering use. In: Produc-
    tion d'agents thérapeutiques par génie génétique. Joyeaux A, Leygue G, Morre M, Roncucci R,
    Schmelck PH (eds) Symposium Quo Vadis? 29-30ᵗʰ May 1985. Toulouse-Labège. France,
    p 45
37. Gene Transfer (1986) Kucherlapati R (ed) Plenum, New York
38. Goldberg AM (1987) The use of animal cells for evaluation of toxicity, carcinogenesis and
    mutagenesis. In: Spier RE, Griffiths JB (eds) Modern approaches to animal cell technology,
    1st edn. p 747 Butterworths, UK
39. Gomes M, Schmitt D, Dezutter-Dambuyant C, Capara JD, Thivolet J (1985) Path. Biol. 34:
    157
40. Griffiths JB, McEntee ID, Electricwala A, Atkinson A, Sutton PM, Naish S, Riley PA (1985)
    Develop. biol. Standard 60: 439
41. Griffiths JB (1985) Cell products: an overview. In: Spier RE, Griffiths JB (eds) Animal cell
    biotechnology, vol 2, p 3, Academic, New York
42. Griffiths PR (1983) Science 222: 297
43. Groner Y, Lieman-Hurwitz J, Dafni N, Sherman L, Levanon D, Bernstein Y, Danciger E,
    Elroy-Stein O (1985) Ann. N.Y. Acad. Sci. 450: 133
44. Hanotte O, Dejaiffe C, Dubois D, Vanderpoorten P, Menozzi FD, Miller AOA (1987) Biol.
    Cell 59: 175

45. Harris H, Miller OJ, Klein G, Worst P, Tachibana T (1969) Nature 223: 363
46. Hayflick L, Moorhead PS (1961) Exptl. Cell Res. 25: 585
47. Hayle AJ (1986) Arch Virol. 89: 81
48. Higgins IJ (1987) Development and applications of amperometric biosensors. 18th FEBS Meeting. Ljubljana June 28–July 3
49. Hirtenstein M, Clark J (1980) In: Richards RR, Rajan K (eds), p 97 Pergamon, Oxford, UK
50. Hooper ML (1982) Biochim. Biophys. Acta 651: 85
51. Hope J (1988) Perfusion cultures. In: Miller AOA (ed) Advances in animal cell technology, NATO Advanced Research Workshop. Brussels 21–24 Sept. 1987. Nijhoff Publ. (in preparation)
52. Hu WS, Giard DJ, Wang DIC (1985) Biotech. Bioeng. 27: 1466
53. Hu WS (1983) Doctoral disseration, Dept. Appl. Biol. Sci., M.I.T., Cambridge, MA
54. Hu WS, Meir J, Wang DIC (1985) Biotech. Bioeng. 27: 585
55. Jacobs JP, Jones CM, Baille JP (1980) Nature 227: 168
56. Jacobson BS, Ryan US (1982) Tiss & Cell. 14: 69
57. Jonasson J, Harris H (1977) J. Cell Sci. 24: 255
58. Katayama H, Itami S, Koizumi H, Tsutsui M (1987) J. Invest. Dermatol. 88: 33
59. Kaudewitz P, Ruzicka T, Meurer M, Braun-Falco O (1985) Arch. Dermatol. Res. 277: 444
60. Klein TM, Wolff ED, Wu R, Sanford JC (1987) Nature 327: 70
61. Kotler M, Reuveny S, Mizrahi A, Shahar A (1985) Develop. biol. Standard 60: 255
62. Lawrance SK, Smith CL, Srivastava R, Cantor CR, Weissman SM (1987) Science 235: 1387
63. Lee WH, Bookstein R, Hong F, Young LJ, Shew JY, Lee EYHP (1987) Science 235: 1394
64. Lewin R (1987) Science 236: 31
65. Lewin R (1987) Science 235: 747
66. Lindskog U, Lundgren B, Billig D, Lindner E (1987) Develop. biol. Standard 66: 307
67. Lindner E, Arvidsson AC, Wergeland I, Billig D (1987) Develop. biol. Standard 66: 299
68. Litwin J (1985) Develop. biol. Standard 60: 237
69. Loo DT, Fuquay JI, Rawson CL, Barnes DW (1987) Science 236: 200
70. Lydersen BK, Putnam J, Bognar E, Patterson M, Pugh GG, Noll LA (1985) The use of a ceramic matrix in a large scale cell culture system. In: Feder J, Tolbert WR (eds) Large-scale mammalian cell culture. Academic Press, New York, p 39
71. Mansbridge JN, Nickoloff BJ, Morhenn VB (1987) J. Invest. Dermatol. 88: 602
72. Margel S (1983) Appl. Biochem. Biotechn. 8: 523
73. Maroudas NG (1973) Exptl. Cell Res. 81: 104
74. Meigner B (1978) Develop. biol. Standard 42: 141
75. Merten OW, Palfi GE, Steiner J (1986) Adv. Biotechn. Proc. 6: 111
76. Merten OW, Palfi GE (1987) Biotech. Bioeng. 66: 111
77. Microcarrier cell culture. Principles and methods. Pharmacia Fine Chemicals AB. Technical literature 1981
78. Idem p 103
79. Miller AOA, Miller-Faurès A (1981) Develop. biol. Standard 50: 287
80. Miller-Faurès A, Blave A, Caudron M, Sené CL, Miller AOA (1985) Develop. biol. Standard 60: 209
81. Miller SJO, Henrotte M, Miller AOA (1986) Biotech. Bioeng. 28: 1466
82. Mitchell KJ, Wray W (1979) Exptl. Cell Res. 123: 452
83. Mizrahi VA (1986) Bio/Technology 4: 123
84. Montagnon B, Vincent-Falquet JC, Fanget B (1984) Develop. biol. Standard 55: 37
85. Montagnon BJ (1985) Tropical and Geographical Med. 37: S40
86. Murray A, Szostak J (1988) Pour la Science 123: 60
87. Ng JJY, Crespi CL, Thilly WG (1980) Anal. Biochem. 109: 231
88. Nichols WW, Murphy DG, Cristofalo VJ, Toji LH, Greene AE, Dwight SA (1976) Science 196: 60
89. Nilsson K, Buszaki F, Mosbach K (1986) Bio/Technology 4: 989
90. Nilsson K, Scheirer W, Merten OW, Östberg L, Liehl E, Katinger HWD, Mosbach K (1983) Nature 302: 629
91. Ohashi M, Aizawa S, Ooka H, Ohsawa T, Kaji K, Kondo H, Kobayashi T, Noumura T, Matsuo

M, Mitsui Y, Murota S, Yamamoto K, Itoh H, Shimada H, Utakoji T (1980) Exptl. Geront. 15: 121

92. Olinger JM, Hill DM, Jakobsen RJ, Brody RS (1986) Biochim. Biophys. Acta 869: 89
93. Peters D, Branschomb E, Dean P, Merrill T, Pinkel D, van Dilla M, Gray JW (1985) Cytometry 6: 290
94. Petricciani JC, Salk J, Noguchi PD (1981) Develop. biol. Standard 50: 15
95. Petricciani JC (1985) The use of continuous cell lines in the manufacture of recombinant DNA products. In: Joyeaux A, Leygue G, Morra M, Roncucci R, Schmelck PH (eds) Production d'agents thérapeutiques par génie génétique Symposium Quo Vadis? 29–30ᵗʰ May. Toulouse-Labège. France, p 209
96. Petricciani JC (1987) The liberation of animal cells: psychology of changing attitudes. In: Spier RE, Griffiths JB (eds). Modern approaches to animal cell technology, 1st edn. p 1 Butterworth, UK
97. Pullen KF, Johnson MD, Phillips AW, Ball GD, Finter NB (1985) Develop. biol. Standard 60: 175
98. Rabits CA, Jarrell JA, Abraham EH (1987) J. Biol. Chem. 262: 1352
99. Ratafia M (1987) Bio/Technology 5: 692
100. Reuveny S, Silberstein L, Shahar A, Freeman E, Mizrahi A (1982) In Vitro 18: 92
101. Reuveny S, Mizrahi A, Kotler M, Freeman A (1983) Biotech. Bioeng. 25: 2969
102. Reuveny S, Corett R, Freeman A, Kotler M, Mizrahi A (1985) Develop. biol. Standard 60: 243
103. Revel JP, Nicholson BJ, Yancey SB (1985) Ann. Rev. Physiol. 47: 263
103a. Riordan ML (1987) Bio/Technology 5: 444
104. Rolin-Limbosch S, Moens W, Szpirer C (1986) Carcinogenesis 7: 1235
105. Rupp RG (1985) Use of cellular microencapsulation in large-scale production of monoclonal antibodies. In: Feder J, Tolbert WR (eds) Large-scale mammalian cell culture. Academic Press, New York, p 19
106. Ryan US, Mortara M, Whitaker C (1980) Tiss. Cell 4: 619
107. Sanford KK, Earle WR, Evans VJ, Waltz HK, Shannon JE (1951) J. Nat. Cancer Inst. 11: 773
108. Scaggiante B, Pineschi B, Sustersich M, Andolina M, Agosti E, Romeo D (1987) Transplantation 44: 59
109. Scheynius A, Johansson C, van der Meide PH (1986) Brit. J. Dermatol. 115: 543
110. Schlaeger EJ, Eggimann B, Gast A (1987) Develop. biol. Standard 66: 403
111. Schönherr OT, van Gelder PTJA, van Hees PJ, van Os AMJM, Roelofs HMW (1987) Develop. biol. Standard 66: 211
112. Schulz R, Krafft H, Piehl GW, Lehmann J (1985) Develop. biol. Standard 66: 489
113. Scott FJ (1985) A search for alternatives to animal testing. Biotechnol. Lab. April 4–5
114. Seizinger BR, Rouleau G, Ozelius LJ, Lane AH, St. George-Hyslop P, Huson S, Gusella JF, Martuza RL (1987) Science 236: 317
115. Shahar A, Mizrahi A, Reuveny S, Zinman T, Shainberg A (1985) Develop. biol. Standard 60: 236
116. Sharma SK (1986) Separat Sci. Technol. 21: 701
117. Selden RF, Skokiewicz MJ, Howie KB, Russell PS, Goodman HM (1987) Science 236: 714
118. Shopsis C, Sathe S (1984) Toxicology 29: 195
119. Simhony S, Kosower EM, Katzir A (1986) Appl. Phys. Lett. 49: 253
120. Simhony S, Kosower EM, Katzir A (1987) Biochem. Biophys. Res. Comm. 142: 1059
121. Terasima T, Tolmach LJ (1963) Exptl. Cell Res. 30: 344
122. Thebline D, Harfield J, Hanotte O, Dubois D, Miller AOA (1987) Develop. biol. Standard (submitted for publication)
123. Thivolet J (1986) La peau de remplacement. Pour la Science Mars, p 16
124. Thivolet J, Faure M, Demidem A, Mauduit G (1986) Bull. Acad. Natl. Med. 170: 557
125. Thomson DMP, Phelan K, Scanzano R, Fink A (1982) Intern. J. Cancer 30: 311
126. Tolbert WR, Lewis C Jr, White PJ, Feder J (1985) Perfusion culture systems for production of mammalian cell biomolecules. In: Feder J, Tolbert WR (eds) Large-scale mammalian cell culture. Academic Press, New York, p 97
127. Uren J (1985) The recovery of genetically engineered mammalian proteins. Intern. Biotech. Lab. April, 26
128. Van Brunt J (1986) Bio/Technology 4: 835

129. Van Brunt J (1987) Bio/Technology 5: 199
130. Van Brunt J (1987) Bio/Technology 5: 313
131. Van Brunt J (1987) Bio/Technology 5: 437
132. Van der Ploeg LHT (1987) Separation of chromosome-sized DNA molecules by pulsed field gel electrophoresis. Biotechn. Lab. April, p 8
133. van Wezel AL (1967) Nature 216: 64
134. van Wezel AL (1973) Microcarrier cultures of animal cells. In: Kruse PF, Patterson MK (eds) Tissue Culture: Methods and Applications. Academic Press, New York, p 372
135. Varani J, Dame M, Beals TF, Wass JA (1983) Biotech. Bioeng. 25: 1359
136. Varani J, Dame M, Rediske J, Beals TF, Hillegas W (1985) J. Biol. Standard 13: 67
137. Varani J, Hasday JD, Sitrin RG, Brubaker PG, Hillegas WI (1986) In vitro 22: 575
138. Vitkauskas GV, Canellakis ES (1985) Biochem. Biophys. Acta 823: 19
139. Vlodavsky I, Lui GM, Gospodarowicz D (1982) Cell 19: 607
140. Volc-Platzer B, Leibl H, Luger T, Zohn G Stingl G (1985) J. Invest. Dermatol. 85: 16
141. Wang HY (1984) Biotech. Bioeng. Symp. 14: 601
142. Watt FM (1986) TIBS 11: 482
142a. Waymouth C (1984) Preparation and use of serum-free culture media. In: Sibarsku DW, Sato GH (eds) Cell culture methods for molecular and cell biology, vol 1, p 23, Alan R. Liss, New York
143. Weissman BE, Saxon PJ, Pasquale SR, Jones GR, Geiser AG, Stanbridge EJ (1987) Science 236: 175
144. Whiteside JP, Spier RE (1985) Develop. biol. Standard 60: 305
145. Widell A, Hansson BG, Nordenfelt E (1984) Virol. Meth. 8: 63
146. Wiemann MC, Creswick B, Calabresi P (1985) Cellular and biochemical aspects of human tumor cell growth and function in hollow fiber culture. In: Feder J, Tolbert WR (eds) Large-scale mammalian cell culture. Academic Press, New York, p 125
147. Wiener F, Klein G, Harris H (1971) J. Cell Sci. 8: 681
148. Wilson T (1984) FDA loosens restraints on cell substrates Bio/Technology, October, p 842
149. Wong H, Chang TMS (1986) Intern. J. Artif. Organs 9: 335
150. Young MW, Dean RC (1987) Bio/Technology 5: 835
151. Zieve GW, Turnbull D, Mullins JM, McIntosh JR (1980) Exptl. Cell Res. 126: 397

Synthesis and Application of Water-Soluble Macromolecular Derivatives of the Redox Coenzymes NAD(H), NADP(H) and FAD

Andreas F. Bückmann*[1] and Giacomo Carrea[2]

[1] Department of Enzyme Technology, Gesellschaft für Biotechnologische Forschung mbH (GBF), Mascheroder Weg 1, D-3300 Braunschweig, FRG
[2] Istituto di Chimica Degli Ormoni, CNR, Via Mario Bianco 9, 20131 Milano, Italy

During the past 15 years, the development of strategies to apply the catalytic potential of redox coenzyme-requiring enzymes has been a subject of intensive study. The main purpose of which has been to cut the cost of coenzyme to an economically acceptable level. One approach has been the utilization of isolated coenzyme-dependent enzyme systems with simultaneous enzymatic coenzyme regeneration (recycling). This has been used in conjugation with ultrafiltration reactor technology (enzyme membrane reactor), with coenzyme concentration being kept at a catalytic level. The concept implies confinement (immobilization) and practically 100% retention of both enzymes and coenzymes being dissolved in homogeneous solution within the reactor space that is closed off by an ultrafiltration membrane through which low-molecular-weight reactants (substrates and products) can freely pass. Since the problem of retaining nearly 100% native coenzymes of relatively low molecular weight by ultrafiltration membranes has not been satisfactorily solved, active macromolecular coenzyme derivatives are required. In this review, the syntheses, properties and merits of water-soluble macromolecular derivatives of NAD(H), NADP(H) and FAD are considered with respect to their biotechnological application.

---

* To whom correspondence should be addressed.

Advances in Biochemical Engineering/
Biotechnology, Vol. 39
Managing Editor: A. Fiechter
© Springer-Verlag Berlin Heidelberg 1989

1 Introduction

Enzymes (biocatalysts) are proteins that catalyse virtually all chemical reactions of life processes.

In principle, enzymes are ideal catalysts in vitro and have two main advantages over nonenzymatic catalysts:

1) High reaction selectivity, in many cases combined with stereo- and/or regio-selectivity.
2) Usually, high catalytic rates under mild conditions and normal pressure.

It ought to be stressed that the catalytic activity of enzymes is not necessarily limited to compounds of solely natural origin, as enzymes can recognize and convert identical molecule groups of both natural and, usually with lower rates, synthetic compounds. The utilization of enzymes in biochemical analysis is routine. However, the application of enzymes as catalysts for reactions of synthetic interest in vitro has not been fully exploited. This is due to a limited supply of commercially available enzymes, a difficult access to non-purchasable enzymes of potential interest or, simply, a more or less misplaced reserved attitude towards employing enzymes for preparative purposes.

The main disadvantages related to biocatalysts are:

1) Instability by inactivation and denaturation due to environmental conditions (strong alkaline or acid conditions, presence of organic reagents, high temperature).
2) Inapplicability to many important types of reactions in cases of a too high substrate specificity.
3) Subject to product or substrate inhibition.
4) Limited availability due to low-level expression by the natural source, e.g. microbial cells.
5) Lack of suitable down-stream processing methods for a fast and easy high-yield isolation and purification adapted to large-scale operation.

Recombinant DNA techniques for site-directed mutagenesis and increased expression levels, in combination with the development of effective large-scale down-stream processing, may lead to better accessible, versatile and stable enzymes in the near future [1].

Transformations catalysed by enzymes such as hydrolases [proteases, amidases, lipases, (gluco)-amylases] [2], nitrile converting enzymes [nitrilase and nitrile hydratase] [2, 3] or isomerases [glucose isomerase] [2], which do not require organic coenzymes, are or are becoming progressively more established in biotechnology (food and pharmaceutical industry) and in preparative organic chemistry.

Advances in the field of enzyme immobilization [4] and biocatalysis in systems with low water content [5] have contributed considerably.

On the other hand, the application of organic (heterocyclic) coenzyme-dependent enzymes, particularly, oxidoreductases [NAD(P)(H)-dependent dehydrogenases], ATP-dependent phosphotransferases or synthetases (ligases), and FAD-dependent oxidases, has been more limited in spite of their great potential for the synthesis and modification of fine chemicals of high value such as chiral synthons, steroids, optically pure labeled compounds (e.g. L-amino acids), precursors of pharmaceutical compounds or for the monitoring of biochemical compounds as an essential part of biosensors.

One of the main obstacles for the use of enzymes requiring relatively weakly interacting coenzymes ($K_d > 10^{-5}$ M) has been the high costs of the coenzyme if used as free stoichiometric reactants (NAD 710 \$ Mol^{-1}, NADH 3.050 \$ Mol^{-1}, NADP 25.780 \$ Mol^{-1}, NADPH 216.100 \$ Mol^{-1}, ATP 800 \$ Mol^{-1})[6-7]. For enzymes dependent on, for example, non-covalently bound FAD, the enzyme-coenzyme interaction is usually so strong ($K_d < 10^{-8}$) that the coenzyme can essentially be considered as an enzyme-bound stoichiometric reactant acting as a covalently attached prosthetic group in a holoenzyme and thus needed in catalytic amounts (regeneratable by catalytic oxidation, e.g. by molecular oxygen, thus facilitating the problem of coenzyme regeneration).

However, several effective methods for the continuous in situ regeneration of the active form of NAD(P) (H) and ATP have been developed in the last 15 years such that these coenzymes can also be used repeatedly in catalytic amounts, considerably reducing the overall production costs of the final product.

Combined with simultaneous NAD(P) (H) regeneration, the structural, enantiomeric and/or prochiral specificity has been exploited for nicotinamide coenzyme-requiring dehydrogenases, e.g. alcohol dehydrogenase [NAD(H)-dependent from horse liver [8,9] or NADP(H)-dependent from *Thermoanaerobium brockii*[10], enantioselective oxidation of 1,2-diols and aminoalcohols and in asymmetric reduction of aliphatic or polycyclic aldehydes and ketones], glycerol dehydrogenase[11] [NADH-dependent from *Cellulomonas spec.*, interconversion of glycerol- and dihydroxyacetone analogues], 2-enoate reductase[12] [NADH-dependent from *Clostridium tyrobutyricum*, reduction of compounds with α,β-unsaturated carboxylate anions], L-amino acid dehydrogenases [NADH-dependent L-leucine dehydrogenase from *Bacillus sphaericus*[13], *B. cereus*[14] or thermophile *B. stearothermophilus*[15] and NADH-dependent L-phenylalanine dehydrogenase from *Rhodococcus spec.* M4[16], *Sporosarcina ureae*[17] or *B. sphaericus*[18], reductive amination of respectively branched chain aliphatic and aromatic α-ketocarboxylic acids to the corresponding L-amino acids], α-hydroxy acid dehydrogenase [NADH-dependent D-lactate dehydrogenase from *Lactobacillus confusus*[19], conversion of pyruvate to D-lactate; NADH-dependent D- and L-hydroxyisocaproate dehydrogenase from respectively *Lactobacillus casei*[19] and *Lactobacillus confusus*[19], D- and L-hydroxylation of branched-chain aliphatic α-ketocarboxylic acids], regio- and stereoselective hydroxysteroid dehydrogenases [e.g. NAD(H)- dependent 3α- and 7α-hydroxysteroid dehydrogenase from respectively *Pseudomonas testosteroni* and *Escherichia coli* and NADP(H)-dependent 12α-hydroxysteroid dehydrogenase from *Clostridium group P*[20] for the relevant reduction or oxidation of respectively dehydrocholic- and cholic acid].

Combined with simultaneous ATP regeneration, the phosphorylation and synthesis of intermediates for complex biological molecules on the preparative scale have been achieved with high specificity and yield using ATP-dependent enzymes[21]. Some representative examples are hexokinase from bakers' yeast (glucose-6-phosphate from glucose), glycerokinase also from *Saccharomyces cerevisiae* (phosphorylated analogs of glycerol) and 5-phospho-D-ribosyl-α-1-pyrophosphate (PRPP) synthetase (PRPP from ribose-5-phosphate). Among the flavoprotein oxidases, glucose oxidase has been used for various biotechnological applications, e.g. as food antioxidant in combination with the hemeprotein oxidase catalase on the industrial scale [22], in immobilized form as part of glucose biosensor systems [22] or as synthetic biocatalyst in the preparative production of hydroquinone from benzoquinone [23]. Highly expressed L-amino acid oxidase from microbial origin, in combination with catalase as H_2O_2 scavenger, may have potential for the production of α-keto acids [24,25]. Other flavin-containing enzymes, such as lactate oxidase or D-amino acid oxidase, should be checked for their applicability. Generally, flavin-containing enzymes have the advantage of an easy coenzyme regeneration as long as the flavin does not dissociate from the enzyme. Coenzyme losses during utilization in the long term are not avoidable in the case of flavoprotein enzymes with less strong interaction between enzyme and prosthetic group, as has been reported for D-amino-acid oxidase ($K_D = 0.53 \times 10^{-6}$ M) [26].

NAD(P) (H)-dependent enzymatic syntheses with simultaneous regeneration of the native coenzyme in conjunction with the batch reactor concept have been recently reviewed by Jones [9] and Keinan et al. [10].

Apart from some preliminary studies reporting the utilization of macromolecular ATP derivatives [27,28], ATP-dependent enzymatic syntheses with simultaneous ATP regeneration have been predominantly carried out in batch processes with homogeneously dissolved or heterogeneously immobilized enzymes using native ATP [21].

This review will consider in detail the synthesis, properties and merits of water-soluble macromolecular derivatives of the redox coenzymes NAD(H), NADP(H) and FAD, emphasizing their application to biocatalytic reactor methods based on ultrafiltration techniques for continuous coenzyme-dependent enzymatic syntheses with simultaneous in situ enzymatic coenzyme regeneration.

2 Regeneration of Macromolecular Coenzyme Derivatives

2.1 General Aspects

The systems developed so far for the in situ regeneration of coenzymes have been reviewed in detail by several authors, e.g. Wang and King [29] [NAD(P) (H), ATP and FAD] and, recently, Crans et al. [21] [ATP], Lee and Whitesides [30] and Chenault and Whitesides [6] [NAD(P) (H), comparing enzymatic and (electro)chemical methods]. Obviously, the non-selectivity of (electro)chemical regeneration methods gives rise to such a substantial formation of inactive coenzyme byproducts that an efficient coenzyme utilization in the long term is not guaranteed. Furthermore, enzymes catalysing product formation may be withdrawn from the system by denaturation on the surface of the electrode or by deactivation by the redox compounds in the case of electrochemical or chemical coenzyme regeneration methods. Product purification may be complicated, especially, if poisonous compounds are present, e.g. methyl-viologen or similar derivatives.

Enzymatic methods for coenzyme regeneration should be preferred, because of high reaction rates, reaction selectivity and the compatibility with enzyme-catalysed synthesis.

The reduction to an insignificant level of the coenzyme costs compared to the overall costs of product synthesis is the main purpose of coenzyme regeneration.

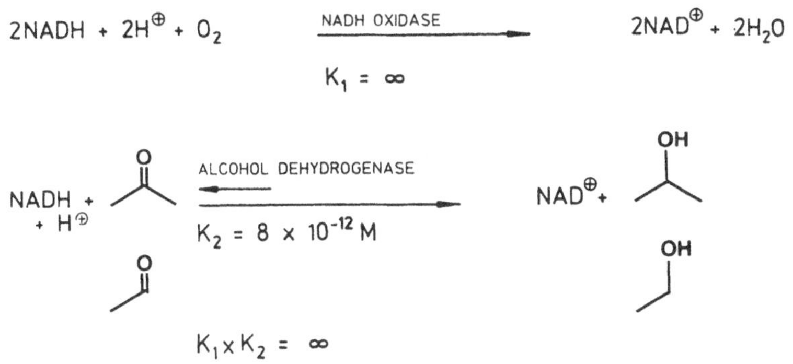

$$2 NADH + 2H^{\oplus} + O_2 \xrightarrow{\text{NADH OXIDASE}} 2 NAD^{\oplus} + 2H_2O$$

$$K_1 = \infty$$

$$K_1 \times K_2 = \infty$$

Fig. 1. Shift of the oxidation of alcohol catalysed by NAD-dependent alcohol dehydrogenase towards ketone/aldehyde formation by NADH oxidase

This can be realized by inserting an enzyme-catalysed reaction that can repeatedly regenerate (recycle) the used coenzyme in the active form. In this way, the total amount of coenzyme continuously present in the enzyme system can be reduced to a catalytic level. Simultaneously, the accumulation of used coenzyme product that may cause product inhibition can be prevented. The low non-stoichiometric coenzyme concentration may simplify the purification of the product, especially for batch processes. Additionally, the coenzyme regeneration reaction can be used to drive the position of the all-over equilibrium of a coupled enzymatic reaction system towards product formation, as has been illustrated in Fig. 1. In this example, NADH oxidation catalysed by NADH-oxidase drives the thermodynamically unfavorable enzymatic formation of NADH and ketone or aldehyde catalysed by alcohol dehydrogenase [31].

Chenault and Whitesides [6] considered the following criteria for an ideal enzymatic coenzyme regeneration method:

1) Inexpensive, readily available and stable enzymes.
2) High specific activity.
3) Simple and inexpensive reagents (substrates and products) that do not interfer with the isolation of the product of interest and cause no deterioration of the coenzymes and enzymes.
4) Selective and high-yield coenzyme regeneration reaction for maintaining functional coenzyme stability.
5) Conditions for maintaining chemical coenzyme stability.
6) High turnover number TN (=number of redox cycles per molecule of coenzyme per unit of time = moles product formed per mol coenzyme per unit of time) representing the regeneration capacity of the system and having a direct impact on the production capacity of coupled enzyme systems.
7) High total turnover number TTN (=mol per product formed per mol coenzyme used during the reaction period) reflecting the coenzyme loss and, consequently, the coenzyme costs.
8) An overall equilibrium for the coupled enzyme system favorable for product formation.

9) No interference with the method of analysis for determining the extent of transformation.

Since all these criteria cannot be fulfilled by any method of enzymatic coenzyme
regeneration, the best compromise ought to be found by selecting a method depending
on the enzyme system chosen.

In this review only those enzymatic coenzyme regeneration methods for the
nicotinamide-containing coenzymes NAD(H) and NADP(H) will be considered
in detail, that are established or of potential use in connection with reactor concepts
based on ultrafiltration, e.g. the enzyme membrane reactor concept (Sects. 2.2 and
2.3).

The application of ultrafiltration reactors with coenzyme-dependent enzyme
systems requires the full retention of both enzymes and coenzymes by the ultrafiltration membrane. The development of ultrafiltration membranes giving acceptable
retention of native coenzyme and unhindered passage for low-molecular-weight
substrates and products has led to encouraging, but not yet totally satisfactory
results in the case of NAD(H) [32] and NADP(H) [33, 34]. For this reason, macromolecular coenzyme derivatives have been considered for utilization in nicotinamide-
containing coenzyme-dependent enzyme systems used in ultrafiltration membrane
reactors. This implies additional criteria as prerequisites for an effectively functioning
system:

I) Approximately 100% retention of enzymes and macromolecular coenzyme
   derivatives by the ultrafiltration membrane of the reactor system.
II) Maintainance of a coenzyme activity for the macromolecular coenzyme derivatives in both reduced and oxidized form of similar order compared to the native
   coenzymes (similar K_M and V_{max}).
III) Simple and high-yielding synthetic procedures to prepare macromolecular
   coenzyme derivatives of well-defined uniform character.

2.2 In Situ NAD(P) Regeneration

Three enzymatic regeneration methods for oxidized nicotinamide-containing coenzymes may be applied in ultrafiltration membrane reactor concepts, since the
enzymes involved, glutamate dehydrogenase (bovine liver), L-lactate dehydrogenase
(rabbit muscle) and alcohol dehydrogenase (yeast), recognize macromolecular NAD-
or NADP derivatives as coenzymes quite satisfactorily [35, 36].

Lee and Whitesides [30] have compared several regeneration strategies. They
concluded that the NH_4^+/α-keto-glutarate/glutamate dehydrogenase reaction system
(Fig. 2A) provides the most satisfactory and generally applicable method for the
regeneration of the oxidized nicotinamide coenzymes NAD and NADP for preparative
enzymatic syntheses in batch processes. Although no relevant results have been
reported, there is no reason why this method should not have merits for continuous
syntheses using the enzyme membrane reactor concept.

In fact, most of the criteria for an ideal coenzyme regeneration method are fulfilled
by this regeneration strategy. Any NAD(P)-dependent coupled enzymatic oxidation
reaction will be driven to near completion, because of the high value of the
equilibrium constants of the reactions, 6.5×10^{14} M^{-1} for NAD and 11.6×10^{14} M^{-1}
for NADP.

α-keto-glutarate and NH_4^+ salts are inexpensive compounds, and these and L-glutamate are innocuous to enzymes. Glutamate dehydrogenase from bovine liver is readily available and inexpensive. Depending on the final product of the synthetic reaction, α-keto-glutarate and L-glutamate, present in stoichiometric amounts, may complicate the product purification. The specific activity of bovine liver glutamate dehydrogenase, around $40 \, U \, mg^{-1}$ for NADH oxidation, must be considered as moderate. Consequently, moderate TN values (Sect. 2.1) are to be expected for syntheses using this coenzyme regeneration system.

The pyruvate/L-lactate dehydrogenase reaction system might be an alternative method limited to the regeneration of macromolecular derivatives of NAD. L-lactate dehydrogenase from rabbit muscle is a more stable and less expensive enzyme than glutamate dehydrogenase. Also the specific activity for L-lactate dehydrogenase is much higher ($1.000 \, U \, mg^{-1}$ for NADH oxidation). The end form of pyruvate can react with the oxidized nicotinamide by a nucleophilic addition reaction [37]. Lactate dehydrogenase catalyses this enolization as part of the reaction mechanism, consequently accelerating this addition reaction [6]. A main disadvantage may be the non-selectivity of the regeneration reaction as nicotinamide modification may only become noticeable after long runs, e.g. during the long-term course of a continuous synthesis utilizing an enzyme membrane reactor.

The evaluation of the advantages of substitution of pyruvate by glyoxylate is in progress, as the latter compound is less expensive and cannot enolize [6]. Furthermore, the equilibrium constant for the glyoxalate/NADH reaction is even higher than for the reductive amination of α-keto-glutarate by glutamate dehydrogenase [6]. The

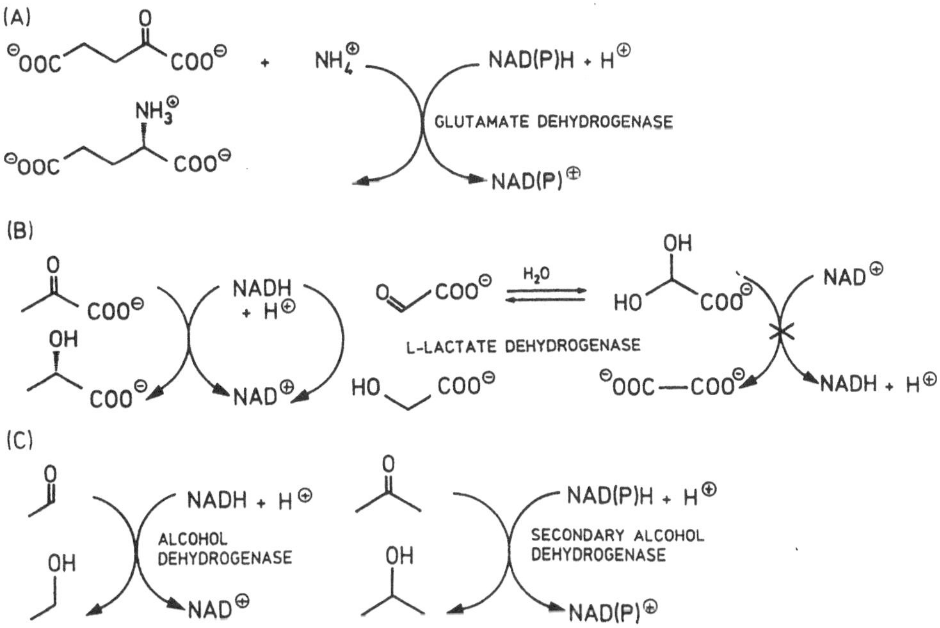

Fig. 2A–C. Enzymatic reactions of potential importance for the regeneration of macromolecular NAD(P) derivatives

NAD-regenerating reactions catalysed by L-lactate dehydrogenase are summarized in Fig. 2B.

The extent of the interference of the NAD-dependent oxidation of glycolate formed by hydration of glyoxal to oxalate in aqueous solution with the NAD-regenerating reduction of glyoxalate to 1-hydroxy-acetate, that are both catalysed by L-lactate dehydrogenase, remains to be investigated. It may be anticipated that the glyoxalate/L-lactate dehydrogenase reaction system for NAD regeneration might successfully be integrated into a continuous operating membrane reactor if the right conditions for substrate input, product output and the mutual concentration of the participating enzymes can be found in order to suppress oxalate formation.

The acetaldehyde/alcohol dehydrogenase enzyme system has been applied for the regeneration of NAD using immobilized, commercially available, cheap alcohol dehydrogenase from yeast (300 U mg^{-1}) for batch NAD-dependent enzymatic synthesis [38].

Acetaldehyde is a low-cost compound and both ethanol and acetaldehyde are volatile enough not to complicate product purification. Ethanol and, especially, reactive acetaldehyde may lead to enzyme deactivation or, in the case of acetaldehyde, to self-condensation and even to nicotinamide modification. In the latter case, the efficiency of the coenzyme utilization may be negatively influenced, as described above for pyruvate.

Although no results have been reported, the acetone/secondary alcohol dehydrogenase enzyme system might be applicable for the regeneration of macromolecular NAD and NADP using respectively secondary alcohol dehydrogenase from *Candida boidinii* (200 U mg^{-1}) [39] and from *Thermoanaerobium brockii* (90 U mg^{-1}) [10].

The various potential reactions for NAD(P) regeneration by alcohol dehydrogenases have been outlined in Fig. 2C.

2.3 In Situ NAD(P)H Regeneration

For batch reactors using native coenzymes, the systems formate/formate dehydrogenase (NADH regeneration) [13], glucose-6-phosphat/glucose-6-phosphate dehydrogenase (NAD(P)H regeneration) [40] and glucose/glucose dehydrogenase (NAD(P)H regeneration) [41] have been intensively investigated.

Only the glucose/glucose dehydrogenase system can not be utilized for the enzyme membrane reactor concept, as macromolecular NAD(P) derivatives, e.g. PEG (M_r 20,000)-N^6-(2-aminoethyl)-NAD(P), were not coenzymatically active with glucose dehydrogenase in the concentration range 0.1–2 mM [42]. It was found that, intrinsically, the adenine modification of the coenzyme prevented an acceptable coenzyme activity with glucose dehydrogenase, as no activity was observed in the same concentration range [42] with low-molecular-weight N^6-(2-aminoethyl)-NAD(P).

If glucose-6-phosphate dehydrogenase from *Leuconostoc mesenteroides* is chosen for the glucose-6-phosphate/glucose-6-phosphate dehydrogenase system, both NADH and NADPH can be regenerated. Wong and Whitesides [40] showed the merits of this enzyme system for the regeneration of native NAD(P)H for batch syntheses of chiral α-hydroxy acids and alcohols on a preparative scale. NAD(P)-specific glucose-6-phosphate dehydrogenase is commercially available and relatively inexpensive.

This enzyme is insensitive to oxidation due to the absence of thiol groups in the active center [6] and thus very stable.

The substrate and products of the regeneration reaction have no adverse effect on most enzymes. Thermodynamically, NAD(P) reduction is strongly favored, as 6-phosphogluconolactone is spontaneously hydrolysed to 6-phosphogluconate, thus making the NAD(P)H formation reaction irreversible.

Glucose-6-phosphate is still an expensive biochemical, although Kazlauskas and Whitesides [43] could reduce the cost to 50–100 \$ Mol^{-1} by developing a batch procedure for the ATP-dependent phosphorylation of glucose at the 6 position by hexokinase, regenerating ATP with the acetylphosphate/acetate system. Similar to inorganic phosphate, glucose-6-phosphate and 6-phosphogluconate catalyse the hydration of the nicotinamide of NAD(P)H, thus inactivating the coenzyme as reducing reagent [6].

As enzymatic synthesis using an enzyme membrane reactor continues over a long period of time, the glucose-6-phosphate/glucose-6-phosphate dehydrogenase system does not show efficient coenzyme utilization.

Glucose-6-sulfate has been proposed as reducing reagent for NAD(P). This compound does not hydrate the C-4-reduced nicotinamide and is prepared more easily than glucose-6-phosphate by a chemical method at approximately the same production costs [44]. Only NADP-dependent yeast glucose-6-phosphate dehydrogenase showed acceptable rates for glucose-6-sulfate as substrate.

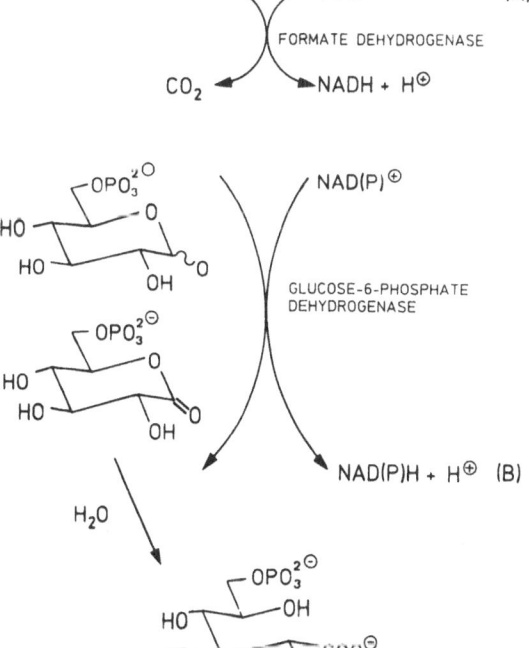

Fig. 3A, B. Enzymatic reactions established or of potential importance for the regeneration of macromolecular NAD(P)H derivatives

The glucose-6-sulfate/yeast glucose-6-phosphate dehydrogenase system may prove to be practical for the continuous synthesis of chemical specialities of very high value by NADPH-dependent enzymes. The glucose-6-phosphate/glucose-6-phosphate dehydrogenase system for NAD(P)H regeneration is outlined in Fig. 3B.

Long-term continuous NADH-dependent enzymatic synthesis of various chiral compounds (L-amino acids, D- and L-α-hydroxy acids) has been successfully carried out on a preparative scale by the membrane reactor concept using the formate/formate dehydrogenase enzyme system for NADH regeneration [45,46]. The formate dehydrogenase-catalysed regeneration of NADH or macromolecular derivatives of NADH is outlined in Fig. 3A ($K = 6.2 \times 10^5$ with the equilibrium on the NADH side).

Suitable intracellular dimeric formate dehydrogenases have been found in methylotrophic yeasts and bacteria. Recently, an excellent bioprocess has been developed for the production of NAD-dependent formate dehydrogenase from the methylotrophic yeast *Candida boidinii* (450.000 U per 5.000 l) [47].

In one operation, affinity partitioning in aqueous two-phase systems using polyethylene glycol ($M_r = 6,000$)-Procion-Red-HE-3B as affinity ligand in the top phase, almost pure formate dehydrogenase (4–6 U mg^{-1}) could be isolated from a crude extract (73% yield, with costs of approx. 1 $ per 1.000 U for the affinity step) [48]. Because of improved access to the enzyme, the formate/formate dehydrogenase system meets almost all the criteria of an ideal method for coenzyme regeneration.

Other advantages are the strong reducing character of the inexpensive reagent formate, driving the NAD reduction practically to completion, the stability of formate and CO_2 and their general innocuousity to enzymes, NAD(H) and derivatives under mild aqueous conditions (pH 6.5–7.5) and the stability of formate dehydrogenase, if autooxidation is prevented.

The low specific activity of the NADH-producing formate dehydrogenase is a serious disadvantage of this coenzyme regenerating system, necessitating a relatively high protein concentration in the closed reactor volume of a membrane reactor to achieve an acceptable TN and TTN value.

Another drawback is the fact that a similar NADP-dependent formate dehydrogenase for NADPH regeneration has not been discovered, e.g. in methylotrophic microorganisms.

It should be noted that NADP-dependent formate dehydrogenases do exist in cellular membranes of strictly anaerobic microorganisms that play a role in the electron transport system, e.g. a selenium-, tungsten- and iron-containing formate dehydrogenase found in *Clostridium thermoaceticum* with a very high specific activity of 1.100 U mg^{-1} [49].

Unfortunately, these formate dehydrogenases are, without exception, extremely sensitive to oxygen due to the presence of Fe—S clusters which are totally absent in the "slow" NAD-requiring formate dehydrogenases from methylotrophs, making the "fast" formate dehydrogenases useless for coenzyme regeneration.

The enzyme systems for NAD(P)H regeneration must be combined with reactions giving NAD(P)H-dependent substrate conversion into product catalysed by a different enzyme.

For alcohol dehydrogenases with a broad substrate specificity that catalyse the

NAD(P)H-dependent reduction of special ketones or aldehydes to the corresponding alcohols, the NAD(P)H regeneration can also be catalysed by the same enzyme in the presence of ethanol or isopropanol in excess depending on the kind of alcohol dehydrogenase.

For this coupled substrate mono-enzyme system with simultaneous NAD(P)H regeneration [10], a main prerequisite is the tolerance to 10–20 % ethanol or isopropanol in aqueous solution by the alcohol dehydrogenase. NAD(H)-dependent primary alcohol dehydrogenase from the thermophile *Sulfolobus solfataricus* (NAD(H)-dependent) [50] and the secondary alcohol dehydrogenases from *Candida boidinii* (NAD(H)-dependent) [39], and *Thermoanaerobium brockii* (NADP(H)-dependent) [10] may be good candidates to use under coupled substrate conditions, e.g. in an enzyme membrane reactor, as macromolecular NAD(P) (H), e.g. polyethylene glycol (M_r = 20,000)-N^6-(2-aminoethyl)-NAD(P), is satisfactorily active with the alcohol dehydrogenases mentioned above (Bückmann and Carrea, unpublished results).

3 Synthesis of Water-Soluble Macromolecular NAD(H), NADP(H) and FAD Derivatives

3.1 General Considerations

A main prerequisite for the application of water-soluble macromolecular coenzyme derivatives in an enzyme membrane reactor is the maintenance of an acceptable coenzyme activity.

This restricts the methods of synthesis, since those parts of the enzyme molecule essential for coenzyme function and interaction with the active sites must be excluded from derivatization. X-ray diffraction studies on the three-dimensional structure of crystalline enzyme complexes with NAD(P) [51] or FAD [52] point in many cases to an outward directed position of the adenine part of the coenzyme molecule embedded in the active site. Preferably, the N^6-amino group and to a lesser extent the C-8 position of the adenine nucleus should be modified, in order to maintain the coenzyme activity of the macromolecular coenzyme derivatives. Mild reaction methods already existed or could be developed for appropriate modifications of the adenine nucleus to introduce functional groups for covalent attachment to suitable water-soluble polymers or copolymerization with various monomers to give macromolecular coenzyme derivatives of NAD(H), NADP(H) and even FAD (Sect. 3.2).

Whether interaction of a macromolecule-bound coenzyme will also result in coenzyme activity or not, depends on the preservation of the right orientation within the active center.

Unfavorable properties of the backbone of the water-soluble macromolecule due to charge, hydrophobicity, hydrophilicity or, simply, the bulky character, may restrict the spatial accessibility of the coupled coenzyme to the active site. Furthermore, the intrinsic chemical modification of the adenine may prevent a proper orientation of the adenine ring within the active site leading to low active (high apparent K_M) or inactive macromolecular coenzyme derivatives [42]. By inserting a spacer molecule of sufficient length between the coenzyme and the polymer backbone, the enzyme-coenzyme interaction may be facilitated [57].

For the design of a favorable method for the synthesis of macromolecular coenzyme derivatives, the uniformity of the attached coenzyme and the stability of the linkage between the coenzyme derivative and the polymer should be emphasized as criteria. In other words, the macromolecular coenzyme derivatives should be chemically well-defined without any ambiguity concerning the group through which coenzyme attachment occurs. The linkage between the coenzyme and the polymer should be stable under the mild conditions (pH range 6–9, T = 20–37 °C) where coenzyme dependent enzymes usually show optimal catalytic activity.

This has been achieved by first synthesizing a well-defined low-molecular-weight coenzyme derivative with or without a spacer, containing a functional group at the adenine part of the molecule (N^6 or C-8 position) with a reactivity under the coupling conditions chosen that cannot be matched by other groups of the coenzyme derivative. By coupling the functionalized coenzyme derivative to reactive groups present on suitable water-soluble polymers, macromolecular coenzyme derivatives can be prepared (Sects. 3.2 and 3.3). An interesting variation of this approach is the co-polymerization of well-defined vinyl derivatives of coenzymes with vinyl-monomers [54].

Mansson and Mosbach [53] have introduced the expression preassembly method for this approach.

To accomplish the criteria of the preassembly method the purification of the required functionalized coenzymes, including intermediate derivatives as products of the various steps of the synthetic procedure, will be the main operation that may lead to considerable overall loss of material. Particularly, if larger quantities are to be prepared, the pure preassembly method is laborious and time-consuming.

However, the search for approaches to simplify the procedure for the preparation of macromolecular coenzyme derivatives by limiting the synthesis steps and adaption to large-scale preparation has been successful (Sects. 3.3.1 and 3.3.2).

3.2 Chemical Modification of NAD(H), NADP(H) and FAD

During the last 15 years most interest has been shown in the development of methods to synthesize macromolecular NAD(H) derivatives, as relatively more NAD(H)-dependent enzymes have been considered for biotechnological applications than NADP(H)- and FAD-dependent enzymes. Factors in favor of the former type of enzymes have been the better commercial availability or direct accessibility (mostly from microbial source), the broader reaction spectrum of synthetic interest and the lower price of NAD. Generally, the experience gained from the modification of NAD could easily be transmitted to analogous derivatizations of NADP(H). The bonding between the phosphate group and the C-2 position of the ribose of the adenosine part of the molecule remains intact during the usual synthetic procedures. Chemical steps involving carbodiimide coupling reagents should be avoided in the case of NADP analogues due to the activation of the 2-phosphate by these reagents that leads to 2',3'-cyclization at the ribose and, consequently, to inactivation of the coenzyme [55]. An established strategy for the synthesis of functionalized NAD(H), NADP(H) or FAD follows the typical procedural pattern:

1) Alkylation of the N(1) position of the adenine ring to introduce a reactive group (carboxyl-, epoxy-, or amino group), leading to N(1)-functionalized NAD, NADP or FAD.

2) For NAD(P) derivatives: Specific chemical reduction with sodium dithionite $(Na_2S_2O_4)$ of the C-4 position of the nicotinamide of N(1)-functionalized NAD(P) leading to N(1)-functionalized NAD(P)H to achieve chemical stability under the alkaline conditions of the Dimroth rearrangement for the C—N bond between the ribose and the nicotinamide.

3) Dimroth rearrangement of N(1)-functionalized NADH, NADPH or FAD to N^6-functionalized NADH, NADPH or FAD under harsh alkaline conditions in aqueous solution (e.g., pH 10.5–11, 65–70 °C, 2 h) to obtain a coenzymatically more active and chemically stable coenzyme derivative.

4) For NAD(P) derivatives: Optional chemical or enzymatic oxidations of N^6-functionalized NAD(P)H to N^6-functionalized NAD(P) by respectively oxidized riboflavin or FMN, simultaneously recycled by oxygen [56], or, e.g. by alcohol dehydrogenase [54] and glutamatic dehydrogenase [30] using acetaldehyde and NH_4^+/α-ketoglutarate as substrates for respectively NADH and NADPH derivatives.

5) Further derivatization to introduce other functional groups.

N^6-functionalized coenzymes are generally obtained with relatively moderate overall yields (10–30 % range) due to losses during purification operations after each step.

These steps are part of the usual preassembly methods for the preparation of macromolecular coenzyme derivatives [53].

The initial alkylation has been carried out, mainly, with iodoacetic acid [57, 58], propiolactone [59, 60], 3.4-epoxybutyric acid [61, 62], introducing functional carboxyl groups, or diepoxy compounds such as 1, 2, 7, 8-diepoxy octane [64] and ethyleneimine [64–66] for the introduction of epoxy and primary amino groups, respectively. Up to 80 % conversion has generally been achieved. In the case of propiolactone and ethyleneimine, the alkylation reactions have only recently been optimized by decreasing the pH from the usual range 4.5–6.0 to 3.2–3.5, thus reducing the formation of byproducts in favor of the main products N(1)-(2-carboxyethyl-NAD [67], and N(1)-(2-aminoethyl)-NAD and -NADP [68], to give maximal conversions in the range 65–75 %.

The conditions for the specific chemical reductions of the nicotinamide and the Dimroth rearrangement have mostly been carried out as described by Lindberg et al. [57]. N^6-carboxyalkyl-NAD(P) derivatives activated by a water-soluble carbodiimide have been further derivatized by reacting with, e.g. 6-diamino-hexane [57], to introduce a spacered functional primary amino group, anticipating an improved coenzyme activity for the macromolecular coenzyme derivative, or by reacting with compounds having a primary amino and vinyl group, e.g. N^6-acryloyl-L-lysine-methylester [59], to prepare copolymerizable NAD.

In Fig. 4 the traditional N(1) and N^6 functionalizations of adenine-containing coenzymes have been outlined.

An important characteristic common to these alkylation methods is the stability of the C(6)-N-alkyl bond of the final N^6-functionalized coenzyme derivatives under

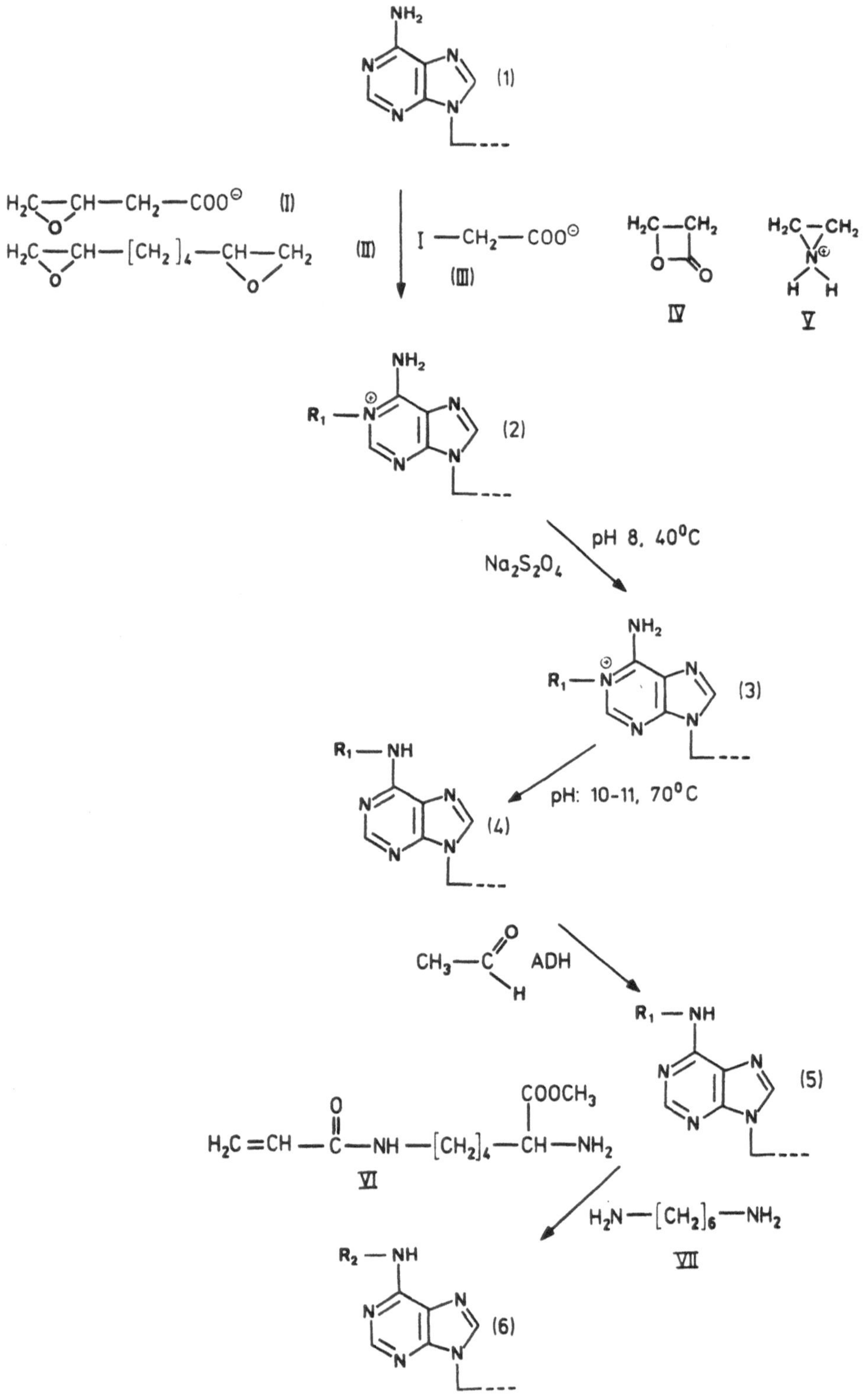

R$_1$ I $^{\ominus}$OOC—CH$_2$—CH—CH$_2$----
                             |
                             OH

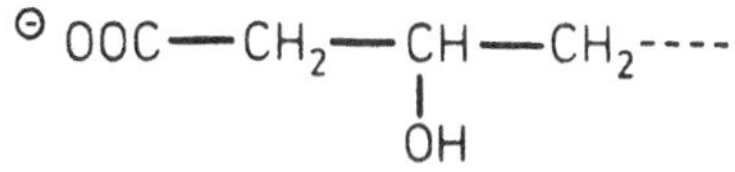

N(1)-AND N^6-(2-HYDROXY-3-CARBOXYPROPYL)-NAD(P)(H) [61,62]

II   H$_2$C—CH—[CH$_2$]$_4$—CH—CH$_2$----
        \/ |
        O   OH

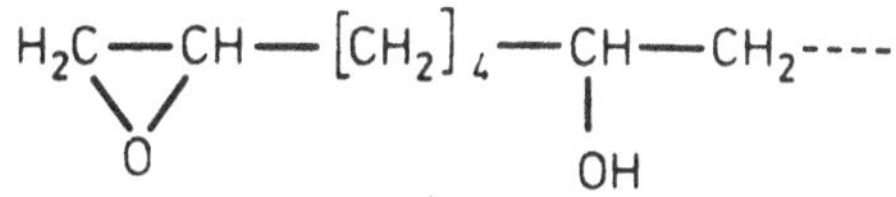

N(1)-AND N^6-(2-HYDROXY-7,8-EPOXY-OCTYL)-NAD(H) [64]

III $^{\ominus}$OOC—CH$_2$----

N(1)-AND N^6-(1-CARBOXYMETHYL)-NAD(P)(H) [57,58]

IV $^{\ominus}$OOC—CH$_2$—CH$_2$----

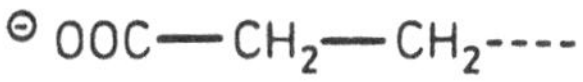

N(1)-AND N^6-(2-CARBOXYETHYL)-NAD(P)(H) [59,60]

V   H$_3$$\overset{\oplus}{N}$—CH$_2$—CH$_2$----

N(1)-AND N^6-(2-AMINOETHYL)-NAD(P)(H) [64-66]

R$_2$ COOCH$_3$
                                    |
     VI   CH$_2$=CH—C—NH—(CH$_2$)$_4$—CH—NH—CO—CH$_2$—CH$_2$----
                   ‖
                   O

N^6-[N-(N-ACRYLOYL-1-METHOXYCARBONYL-5-AMINO-PENTYL)-AMIDOPROPYL]-NAD [59]

VII   H$_3$$\overset{\oplus}{N}$—(CH$_2$)$_6$—NH—CO—CH$_2$----

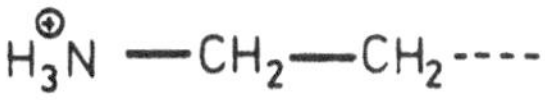

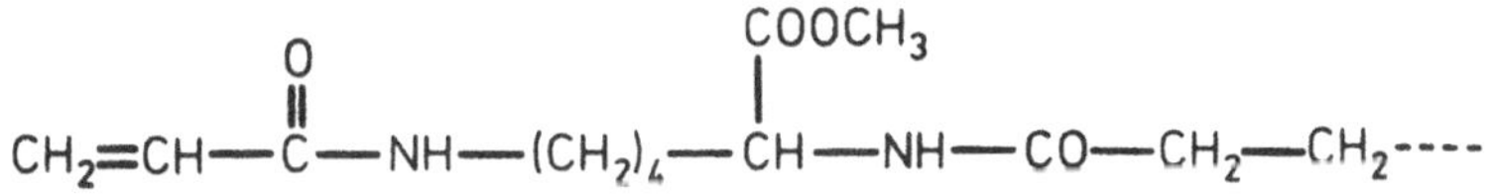

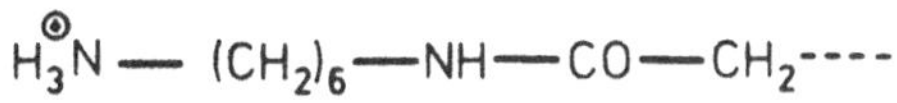

N^6-(6-AMINOHEXYL)-CARBAMOYLMETHYL-NAD(P) [57,58]

Fig. 4. Reaction pathway for the classic N(1)- and N^6-functionalization of NAD(P)(H) and FAD by alkylation N. B. For NAD(P)(H): (3) and (4) are NAD(P)H derivatives; for FAD: steps (1) → (2) → (4) apply in the case of functionalization with I; for II and VI: no NADP derivatives have been synthesized

the usual mild operating conditions for coenzyme dependent enzymes (aqueous solution pH range 6–9). The presence of more than one methylene group in the alkyl part of the functional group introduced is an advantage from the chemical point of view, since only the N^6-carboxymethyl group is unstable under slightly acid conditions, e.g. pH 4.5–4.7, that are optimal for modifications using carbodiimide coupling reagents [69].

As the application of coenzymes with a functionalized adenine nucleus became a subject of study, other modification procedures have been found that follow a different strategy to that described above.

One of the first papers on the synthesis of NAD analogs for the preparation of coenzymically active macromolecular NAD derivatives was by Wykes et al. [70] who describe a method for the direct synthesis of N^6-succinyl-NAD by reacting NAD with succinic anhydride. The advantage of this one step procedure cannot be matched by the disadvantage of the intrinsic instability of the C(6)-N-acyl bond under the moderate alkaline conditions optimal for the catalytic activity of NAD(H)-dependent enzyme systems.

C(8)-functionalized NAD and NADP have been synthesized by Lee and Kaplan [71] for application in affinity chromatography. Bromination with Br_2 of the C(8) position of the adenine in aqueous solution followed by specific chemical reduction with $Na_2S_2O_4$ to C(8)-brominated NADH or NADPH and nucleophilic displacement of Br with 1,6-diaminohexane, introducing a spacered primary amino group, and enzymatic oxidation is similar in length to the traditional preparation of N^6-alkylated-NAD(P). Similarly, Zappelli et al. [72,73] introduced a carboxyl function at the C(8) position of the adenine of NAD(P) by reacting 8-bromo-adenine NAD(P) with the disodium salt of 3-mercaptopropionic acid in dimethyl sulfoxide without previous reduction, to give 8-(2-carboxyethylthio)-adenine-NAD(P).

C(8)-functionalized coenzymes can generally be obtained in good overall yield (60–70%) except for 8-(2-carboxyethylthio)-adenine-NADP (25%) [73].

Compared to native NAD and NADP, N^6-functionalized- and C(8)-functionalized NAD and NADP generally maintain acceptable coenzyme activities with respect to the usual dehydrogenases from different sources such as alcohol dehydrogenases, lactate dehydrogenases, malate dehydrogenases, glucose-6-phosphat dehydrogenases, glutamate dehydrogenases, etc. [42,71–74]. If covalently attached to a polymer, the active site seems to be less accessible for macromolecular C(8)-functionalized coenzymes derivatives than for polymer-bound N^6-functionalized coenzymes, resulting in significant higher apparent K_M and lower V_{max} values [72,73,75] (Sect. 4.3).

The pathways to the other coenzyme functionalizations, developed in parallel with the traditional alkylation methods, have been outlined in Fig. 5.

Recently, a simplified preassembly method has been developed by Bückmann [76,137] for the preparation of macromolecular NAD, NADP, and FAD derivatives starting with N(1)-(2-aminoethyl)-adenine derivatives of these coenzymes, but omitting the specific chemical reduction with $Na_2S_2O_4$ and avoiding the usual harsh alkaline conditions of the Dimroth rearrangement. A striking feature of this method is the different reaction behavior of N(1)-(2-aminoethyl)-NAD, -NADP and -FAD with respect to the Dimroth rearrangement to the corresponding N^6-(2-aminoethyl)-derivatives. Unexpectedly, it was discovered that this rearrangement can still be carried out at a relatively fast rate, but now under unusually mild aqueous conditions

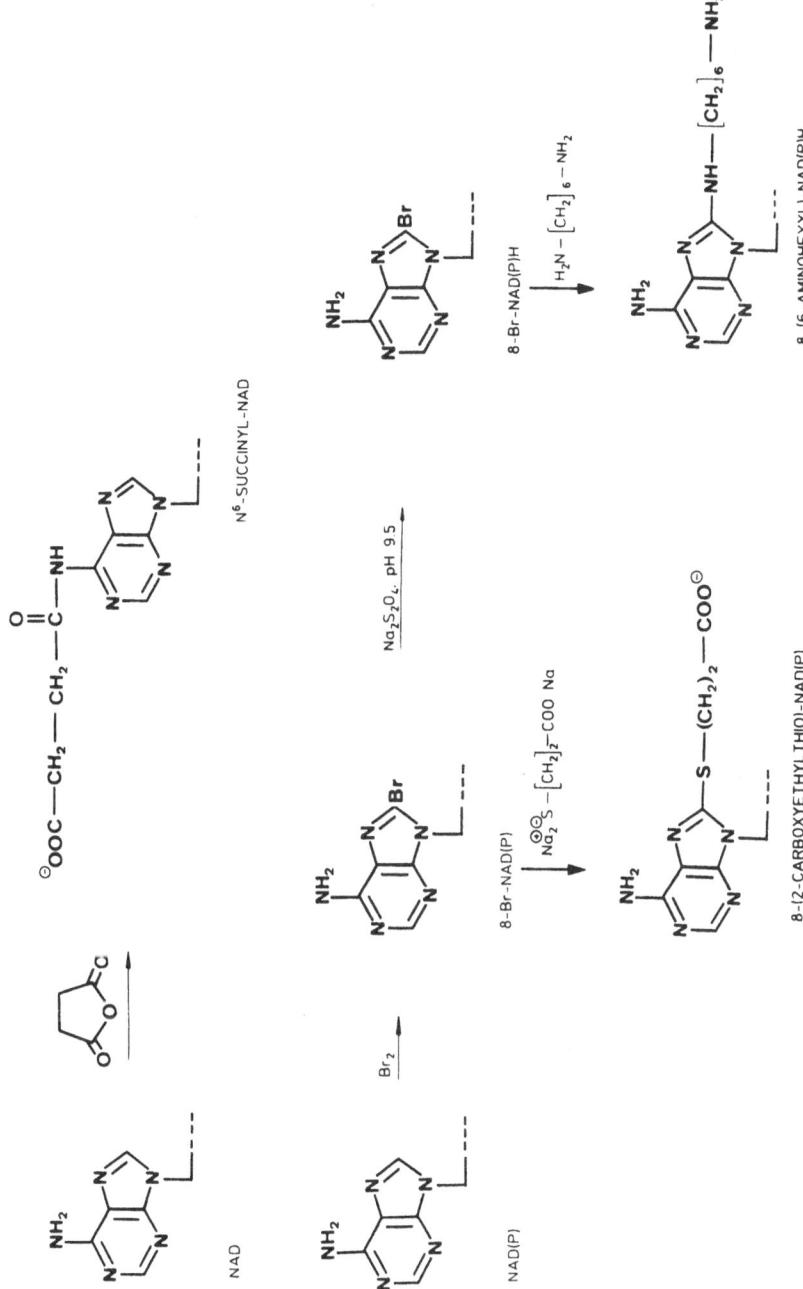

Fig. 5. Reaction pathway for the alternative classic N^6- and C(8) functionalization of NAD(P) (H)

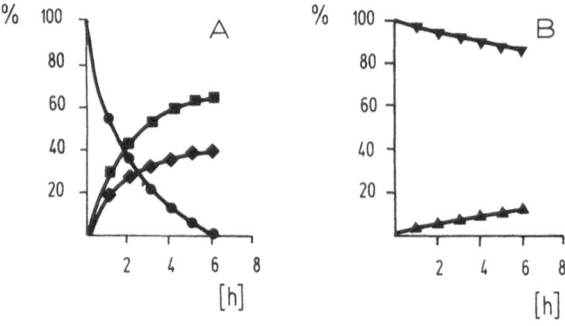

Fig. 6A, B. Comparison of the transformation (%) of N(1)-(2-aminoethyl)-NAD (A) and N(1)-[N-(2-carboxy-propioamido)-ethyl]-NAD (B) under mild aqueous conditions: distilled water, pH 7.0, 50 °C, initial concentration 15 mM. A: (●) N(1)-(2-aminoethyl)-NAD; (■) N^6-(2-aminoethyl)-NAD; (◆) 1,N^6-ethanoadenine-NAD. B: (▼) N(1)-[N-(2-carboxy-propioamido)-ethyl]-NAD; (▲) N^6-[N-(2-carboxy-propioamido)-ethyl]-NAD

(pH 6.0–6.5, 40–50 °C, 4–7 h) with the consequence that no previous reduction of the C-4 position of the nicotinamide is required. Self-evidently, the optional oxidation step can be avoided, too.

This particular reaction behavior is an exception that applies only to N(1)-(2-amino-ethyl)-adenine derivatives. Normal Dimroth rearrangement character (slow transformation under mild conditions) returns, as has been observed for N(1)-[N-(2-carboxy-propio-amido)-ethyl]-NAD synthesized by chemical modification of the aminoethyl group of N(1)-(2-aminoethyl)-NAD with succinic anhydride (Fig. 6) [137]. Another striking aspect is the lower conversion of N(1)-(2-aminoethyl)-NADP to N^6-(2-aminoethyl)-NADP (max. 25–30%) obtained, compared to the corresponding transformation of N(1)-(2-aminoethyl)-NAD and -FAD (max. 60–65%). It has been shown that a parallel transformation to tricyclic 1,N^6-ethano-adenine-NAD, -NADP and -FAD occurs, simultaneously with the Dimroth rearrangement under mild conditions (Fig. 7). The 2-phosphate group of the ribose of N(1)-(2-aminoethyl)-NADP has a similar catalytic effect as inorganic phosphate on the tricyclization reaction of N(1)-(2-aminoethyl)-NAD, producing 1,N^6-ethanoadenine-NADP as the major product and decreasing the transformation yield for N^6-(2-aminoethyl)-NADP [77].

The simplified procedure to synthesize N^6-(2-aminoethyl)-adenine analogues of NAD, NADP und FAD can now be limited to the following steps:

1) Alkylation of NAD, NADP or FAD with ethyleneimine under optimized conditions to N(1)-(2-aminoethyl)-NAD, -NADP or -FAD, without purification of the main product.

2) Incubation of the reaction mixture (typical compositions: 20–25% NAD, NADP or FAD, 65–75% N(1)-(2-aminoethyl)-NAD, -NADP or -FAD, 5–10% N(1)-alkylated byproduct of NAD, NADP or FAD) at pH 6.0–6.5, 40–50 °C, until N(1)-(2-aminoethyl)-NAD, -NADP or -FAD has disappeared (checked by thin-layer chromatography).

3) Purification of N^6-(2-aminoethyl)-NAD, -NADP or -FAD from the reaction mixture after step 2. [typical composition: 40–45% N^6-(2-aminoethyl)-NAD or -FAD or 25–30% N^6-(2-aminoethyl)-NADP, 25–30% 1,N^6-ethanoadenine-NAD

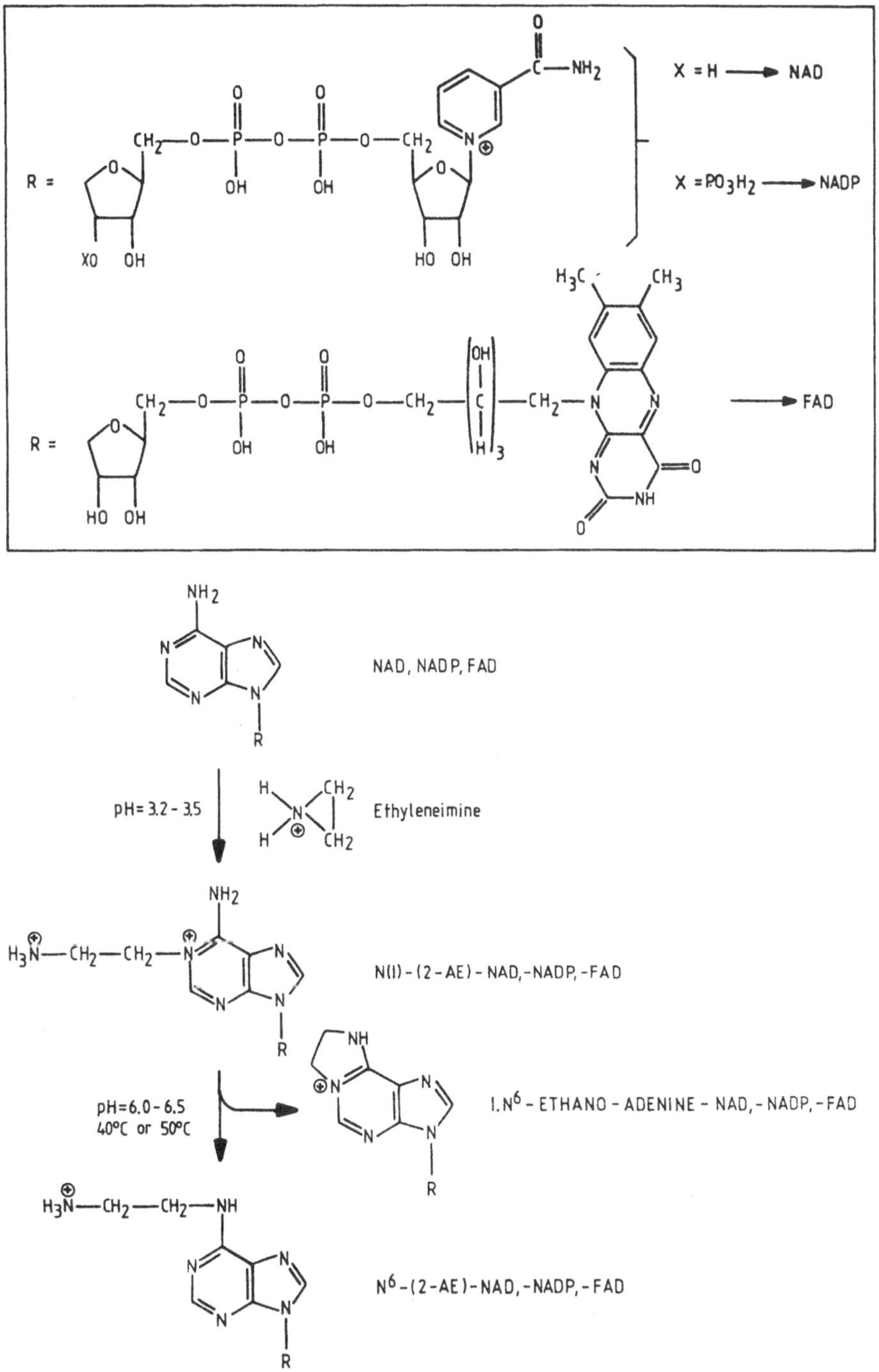

Fig. 7. Reaction pathway for the simplified synthesis of N^6-(2-aminoethyl)-NAD, -NADP and -FAD

or -FAD or 40–50% 1,N⁶-ethanoadenine NADP (% native coenzyme and byproducts see step 2.)] by ionexchange.

The overall yields of pure N⁶-(2-aminoethyl)-NAD, -NADP or -FAD approach the corresponding final composition percentages of the reaction mixture of step 3 [76].

This means that by this simplified method higher overall yields (around 40%) can be achieved for N⁶-(2-aminoethyl)-NAD and -FAD than has been reported for N⁶-functionalized -NAD and -FAD synthesized according to the usual strategies that include intermediate purification operations (yields 10–30%) [59,63].

The simple procedure is currently under investigation with the aim of developing an operational strategy for scale-up.

It should be noted that Sakamoto et al. [67] recently reported an overall yield of 50% for N⁶-(2-carboxyethyl)-NAD prepared according to the traditional procedure from N(1)-(2-carboxyethyl)-NAD, synthesized by an optimized alkylation reaction with propiolactone.

The simplified procedure for the synthesis of N⁶-(2-aminoethyl)-NAD, -NADP and -FAD has been outlined in Fig. 7.

3.3 Water-Soluble Macromolecular Derivatives of NAD(H), NADP(H) and FAD

Having described the functionalization of native NAD(H), NADP(H) and FAD, resulting in low-molecular-weight derivatives with more or less spacered reactive epoxy-, carboxyl- or primary amino groups, various strategies will now be considered for the preparation of water-soluble macromolecular derivatives of these coenzymes, stressing both their advantages and disadvantages.

3.3.1 The Preassembly Strategy

The preassembly strategy [53] for the preparation of water-soluble macromolecular coenzyme derivatives consists of covalently coupling a defined functionalized coenzyme derivative to relevant reactive groups of a water-soluble polymer or copolymerizing suitable coenzyme derivatives with appropriate monomers.

Reactive groups can generally be introduced in the polymer by simple chemical modification reactions or the polymer is directly commercially available in the functional form.

Depending on the polymer chosen, the amount of functional groups per polymer molecule and, consequently, the ratio coenzyme molecules/polymer molecule can be varied. With respect to preassembly methods, the following reactive water-soluble polymers have been utilized for the preparation of macromolecular coenzyme derivatives: cyanogenbromide (CNBr)-activated dextran (M_r 40,000) [78], cyanuric chloride activated dextran (M_r 110,000) [79], glycylglycylglycyl-dextran (M_r 70,000) [80], carbodiimide-activated carboxymethyl dextran (M_r 40,000–500,000) [81], aminated polyethylene glycol (PEG) (M_r 3,000–4,500, 20,000) [60,82,83], carbodiimide or N-hydroxy-succinimide-activated carboxylated PEG (M_r 10,000–20,000) [81,137], polyethylene imine (PEI) (M_r 40,000–60,000) [61–63], polylysine (M_r 3,000–85,000) [84], caseine (M_r 24,000–67,000) [85], carbodiimide activated carboxylated polyvinylpyrrolidone

(M_r 160,000) [137] and various maleic-anhydride-containing copolymers (M_r 18,000–20,000) [137].

It is obvious, that a very limited reaction range is of practical interest for the assembly of functionalized coenzyme analogs and reactive polymers. Coenzyme analogs with primary amino groups are usually coupled to polymers with carboxyl groups activated by a water-soluble carbodiimide coupling reagent, e.g. 1-(3-di-methylaminopropyl)-3-ethyl-carbodiimide-HCl (EDC), under moderate acid conditions (pH 4.5–4.8), or by N-hydroxy-succinimide under weakly alkaline conditions (pH 7.0–8.0), if the polymer is also soluble in nonaqueous solution (e.g. carboxylated PEG in ethylacetate) [86]. Both differently activated carboxyl groups lead to the formation of amide bonds. It should be emphasized that the N-hydroxy-succinimide activation must be carried out for the coupling of NADP derivatives with primary amino groups to avoid the 2′, 3′-cyclization of the 2′-phosphate at the ribose (Sect. 3.1).

Carboxylated coenzyme derivatives are generally coupled to polymers with primary amino groups, e.g. polyethylene imine, polylysine or dextran, modified with di-aminohexane, by the carbodiimide method, as these coenzyme analogs are not soluble under the nonaqueous conditions required for the N-hydroxy-succinimide activation.

Bückmann et al. [81] reported that diamino PEG (M_r 10,000) reacted in low yield with EDC activated N(1)-(1-carboxymethyl)-NAD under equimolar conditions at high concentration. On the contrary, if N(1)-(2-aminoethyl)-NAD was coupled under similar conditions to EDC activated dicarboxyl PEG (M_r 10,000), a high coupling yield was obtained (approx. 90 %).

It should be noted that the strategy to attach carbodiimide activated carboxylated coenzyme derivatives to aminated polymers has two main disadvantages:
1) Acceptable covalent attachment only proceeds in the presence of a molecular excess of polymer bound primary amino groups.
2) The strategy is not satisfactory for coupling carboxylated NADP derivatives (2′, 3′-cyclization reaction).

Both disadvantages can be found in the laborious procedure of Okuda et al. [60] for preparing PEG (M_r 3,000) N^6-(2-carboxyethyl)-NADP, which requires an impractical ionexchange step to remove the excess of diamino PEG and an enzymatic step utilizing 2′, 3′ cyclic-nucleotide-3′-phospho-diesterase to hydrolyse unspecifically the cyclic bound phosphate. A mixture of 2′- and 3′-phosphorylated NADP derivatives was obtained.

Choosing polyethylene imine with a high excess of primary amino groups, Zappelli et al. [62] coupled N^6- and C(8)-carboxylated NADP by the carbodiimide method and obtained limited enzymatically reducible polymer-bound NADP, presumably because of the unnoticed 2′, 3′-cyclization reaction.

In a special case, the chemical coupling step could be replaced by an enzyme catalysed amide bond formation. Applying transglutaminase that catalyses a Ca^{2+} dependent acyl-transfer reaction between γ-carboxy amide groups of peptide-bound glutaminyl residues and primary amino groups of a variety of compounds, N^6-[(6-aminohexyl)-carbamoylmethyl]-NAD was covalently attached to casein, leading to casein-NAD derivatives with only 1–2 coenzyme molecules per caseine molecule [85].

By copolymerizing spacered N^6-vinyl derivatives of NAD with low-molecular-weight vinylated monomers, a variety of NAD-containing copolymers have been

prepared by varying polymer length (M_r 1,000–500,000), hydrophobicity or hydrophilicity of the spacer between the polymer backbone and the integrated NAD
analogue, chemical properties of the macromolecule body (charge and again extent of
hydrophobicity and hydrophilicity) and NAD density (=ratio mol NAD per mol
monomer) [59,87].

The laborious efforts of their preparation and characterization were associated
with a systematic study of the relationship between the properties of copolymerized
NAD derivatives and the coenzymic activity.

The total reducibility or oxidizability for not too bulky macromolecular NAD(P) (H)
derivatives is generally good (80–95%) and is characteristic of the preassembly
strategy [137].

3.3.2 Simplified Strategies

For alkylated N^6-functionalized NAD(H) and NADP(H), various methods have
been reported to overcome the laborious character of the usual preassembly methods
by reducing both reaction and purification steps for the intermediate compounds
and end-products.

In this context, some simplified procedures relate to the simultaneous derivatization
and molecular weight enlargement of the coenzyme.

Le Goffic et al. [88] synthesized macromolecular derivatives of NAD (M_r 20,000–
40,000) with high coenzyme load (up to 33%, w/w) by copolymerizing acrylamide
and a crude reaction product containing approx. 90% N(1)-(acryl-2-hydroxypropyl)-
NAD under mild alkaline conditions (pH 7–8, 40 °C). By a slow reaction (72–120 h),
the Dimroth rearrangement was simultaneously carried out, while bound to the
polymer, without previous chemical reduction of the nicotinamide at the C(4)
position.

Maximally 58% of the polymer-bound NAD was coenzymically active, presumably
due to the presence of inactive nucleotides derived from NAD deteriorated during
the long reaction time of the copolymerization and rearrangement reaction.

Fuller et al. [89] obtained coenzymically active macromolecular spacered NADH
derivatives (21%, w/w) after N(1) alkylation of native NAD by the epoxy groups
of an acrylic water-soluble polymer (M_r 37,000) during 8 d at pH 5.0 and 25 °C and
subsequent fast Dimroth rearrangement under the usual harsh alkaline conditions
after chemical reduction, while bound to the polymer. Starting from NADH, a one-step
reaction procedure was developed, since it was found that incubation at pH 10 at
25 °C effected simultaneous coupling by N(1) alkylation and Dimroth rearrangement
leading after 8 d to a macromolecular NADH derivative with 3% (w/w) coenzyme
content [89].

However, at best 60% total enzymatic oxidizability was observed due to unknown
side reactions.

A compromise between the simplified and laborious preassembly methods is the
simplified procedure for the synthesis of polymer-bound NAD(H) developed by
Bückmann [66,81] that has the advantage of resulting in a uniform coupled NAD(H)
analog.

The essential steps of the procedure are synthesis and purification of N(1)-(2-
aminoethyl)-NAD, covalent attachment to carboxylated water-soluble polymers and,

while covalently bound, chemical reduction to N(1)-(2-aminoethyl)-NADH and subsequent Dimroth rearrangement to N^6-(2-aminoethyl)-NADH under harsh alkaline conditions' followed by an optional enzymatic oxidation to N^6-(2-amino-ethyl)-NAD (Fig. 8) [81]. By this procedure, dextran (M_r 40,000, 150,000 and 500,000)- and PEG (M_r 10,000 and 20,000)-N^6-(2-aminoethyl)-NAD(H) have been synthesized. Based on this strategy, a method for the synthesis of technical grade polyethylene glycol (PEG)-(M_r 20,000)-N^6-(2-aminoethyl)-NADH was developed, omitting the purification step for N(1)-(2-aminoethyl)-NAD [90]. A crude reaction mixture con-taining up to 70% N(1)-(2-aminoethyl)-NAD (mol/mol) was used for coupling. By two partition steps in PEG-NADH/Na_2CO_3 aqueous two-phase systems with high partition coefficients (70–100) for PEG (M_r 20,000)-N(1)-(2-aminoethyl)-NADH and -N^6-(2-aminoethyl)-NADH (present in the top phase), before and after the rear-rangement step, nearly all non-coupled NAD and derivatives could be removed by extraction in the Na_2CO_3 bottom phase. The procedure may easily be adapted to large-scale synthesis, since partition phenomena in aqueous two-phase systems are independent of the total volume of the system [91].

A very acceptable overall yield of 33.5% (Table 1) has been reported for the preparation of technical grade PEG (M_r 20,000)-N^6-(2-aminoethyl)-NADH on a pilot scale (approx. 1 kg) [90]. The total reaction pathway of the simplified preparation

Table 1. Summary of the synthesis of technical grade PEG (M_r 20,000)-N^6-(2-AE)-NADH

Step		mmol	Presence of uncoupled NAD(H) and alkylated NAD(H) (mmol)
I	PEG (M_r 20,000)-[NH—CO—CH_2—CH_2—COOH]$_2$ Reaction mixture containing NAD (35 mmol), N(1)-(2AE)-NAD (93.8 mmol), N(1)-NAD by-products (11.2 mmol) EDC, pH 4.6–4.7, room temperature PEG (M_r 20,000)-N(1)-(2AE)-NAD	76.2[a]	63.8[a]
II	Batch treatment with AGW50-X4, pH 1.4, 4 °C PEG (M_r 20,000)-N(1)-(2AE)-NAD	52.15[a]	9.3[a]
III	Chemical reduction $Na_2S_2O_4$, pH 8.0, 45 °C PEG (M_r 20,000)-N(1)-(2AE)-NADH	52.25[b]	9.3[b]
IV	1st PEG (M_r 20,000)-N(1)-(2AE)-NADH/Na_2CO_3 aqueous two-phase system, pH 10, 4 °C PEG (M_r 20,000)-N(1)-(2AE)-NADH	51.15[b]	5[b]
V	Dimroth rearrangement, pH 11, 65 °C PEG (M_r 20,000)-N^6-(2AE)-NADH	50[b]	6.15[b]
VI	2nd PEG (M_r 20,000)-N^6-(2AE)-NADH/Na_2CO_3 aqueous two-phase system, pH 10.5, 4 °C PEG (M_r 20,000)-N^6-(2AE)-NADH	47[b]	0.85[b]

Note. Overall yield 33.5% (I → VI); yield 62% (II → VI); yield 90% (III → VI)
PEG = polyethylene glycol; 2AE = 2-aminoethyl.
[a] Based on $e_{259} = 18.000$ M^{-1} cm^{-1};
[b] based on $e_{340} = 6.220$ M^{-1} cm^{-1}

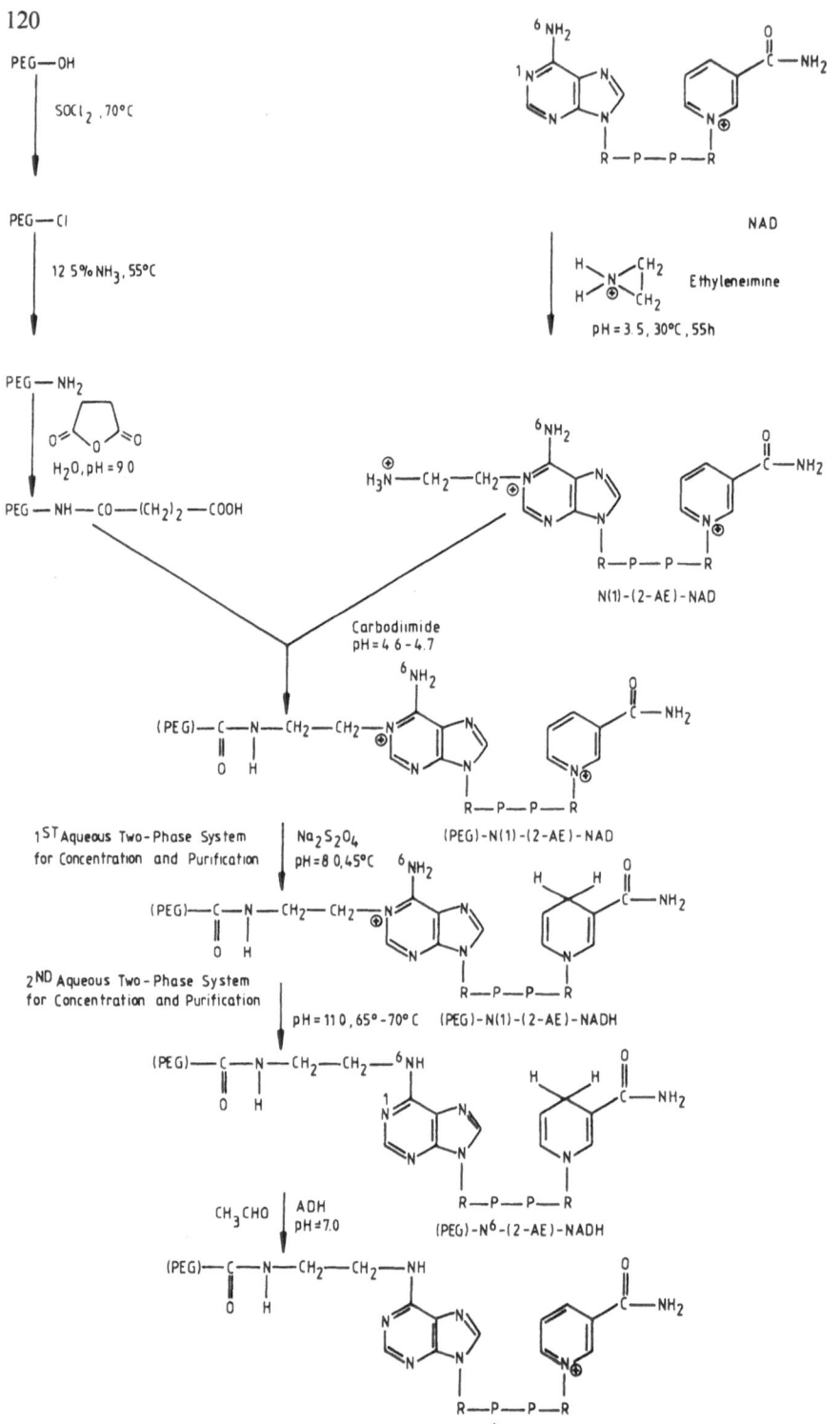

Fig. 8. Reaction pathway for the simplified preparation of (technical grade) PEG-(M_r 20,000)-N^6-(2-aminoethyl)-NAD(H)

of pure and technical grade PEG (M_r 20,000)-N^6-(2-aminoethyl)-NAD(H) is outlined in Fig. 8, including the position of the partition steps. Starting with pure or crude N(1)-(2-aminoethyl)-NAD, high total enzymatic reducability and oxidizability were observed, 90–95 % or 80–85 %, respectively [81,90]. This difference of around 10 % stems from the fact that coenzymically inactive N(1)-NAD byproducts [overreacted N(1)-(2-aminoethyl)-NAD], present in the crude reaction mixture, have also been coupled in the technical grade preparations.

3.4 Water-Soluble Enzyme — Coenzyme Conjugates

An alternative to synthetic polymer-bound coenzymes are enzyme-coenzyme conjugates where the coenzyme is covalently linked to the enzyme through a suitable spacer. The aim is to maintain the interaction between the active site and the enzyme-bound coenzyme. Furthermore, the latter should interact with the active site of another unmodified enzyme with the same coenzyme requirement or react with redox compounds.

This approach was first pursued by Mansson et al. [92] by attaching N^6-[N-(6-aminohexyl)-carbamoylmethyl]-NAD to horse liver alcohol dehydrogenase by a carbodiimide mediated reaction. Since then, several other enzyme-coenzyme conjugates have been synthesized, but this has been restricted to enzyme-NAD conjugates [93–101].

Table 2 shows the enzymes modified, the type of functionalized NAD, the ratio mol NAD per mol subunit, the activity of the enzyme-NAD conjugates in the

Table 2. Characteristics of dehydrogenase-NAD conjugates

Enzyme	NAD derivative	Mol coenzyme / Mol enzyme subunit	Enzyme activity (% of native enzyme)	Cycling rate[a] (% of native enzyme with free NAD)	Refs.
Liver alcohol dehydrogenase	N^6-[N-(6-aminohexyl)-carbamoylmethyl]-NAD	0.9–2.1	40	100[b]	92, 93, 95)
	N^6-[N-(8-amino-3,6-dioxaoctyl)-carbamoylmethyl]-NAD	0.4–0.5	52	27	100)
Lactate dehydrogenase	N^6-[N-(6-aminohexyl)-carbamoylmethyl]-NAD	1.6	28–58		94)
	NAD activated with N^6-attached diazoniumaryl or imidoester groups	0.4–1.1	80–90	45–63[b]	96)
Malate dehydrogenase	Activated PEG (M_r 3,000)-N^6-(2-carboxyethyl)-NAD	0.5–1.2	77–127	< > 100[c]	97)
Glucose dehydrogenase	Activated PEG (M_r 3,000)-N^6-(2-carboxyethyl)-NAD	2.1	92		98)
Formate dehydrogenase	N^6-[N-(6-aminohexyl)-carbamoylmethyl]-NAD	0.2[b]	116	100	99)

[a] The cycling rate [mol product per mol coenzyme (unit of time)$^{-1}$] was determined in coupled-enzyme systems;
[b] calculated taking into consideration only the fraction of bound coenzyme reducable by substrate;
[c] depending on the concentration of the conjugate;
PEG: Polyethyleneglycol

presence of excess of exogenous NAD and the cycling rate of enzyme-linked NAD.

The length of the spacers was about 1 nm except for thermostable malate dehydrogenase-NAD [97] and glucose dehydrogenase-NAD conjugates [98] where the spacer was a linear chain of polyethylene glycol (M_r 3,000) of approx. 25 nm length.

For horse liver alcohol dehydrogenase [92,93,100] and formate dehydrogenase [99], the NAD derivatives were coupled to carboxyl side groups, whereas in other cases primary ε-amino groups of lysine were involved.

In all cases, the linkage between enzyme and coenzyme analogue was found to be stable under the normal moderate alkaline conditions for optimal enzyme activity.

The catalytic integrity of the enzyme protein part of the enzyme-coenzyme conjugates have been tested by measuring relative activities in the presence of excess of free NAD compared to native enzyme. Activity ratios in the range 28–127% have been found (Table 2).

The strategy of previously forming ternary complexes by the dehydrogenase, a pseudosubstrate and the reactive NAD derivative before adding a coupling reagent has been proposed to direct the linking towards amino acid residues in the vicinity of the active site and protecting functional amino acid residues in the active site [92].

This strategy was successfully applied for ternary complexes, e.g. as oxalate-[N^6-(diazonium-aryl)-NAD]-lactate dehydrogenase [101], pyrazole-[N^6-(diazonium aryl)-NAD]-horse liver alcohol dehydrogenase [96] and, recently, pyrazole-[N^6-[N-(8-amino-3,6-dioxaoctyl)-carbamoylmethyl]-NAD]-horse liver alcohol dehydrogenase [100].

Mansson et al. [92] could not bind N^6-[N-(6-aminohexyl)-carbamoylmethyl]-NAD to horse liver alcohol dehydrogenase via the ternary complex with pyrazole. Goulas [100], utilizing the three-dimensional structure of horse liver alcohol dehydrogenase, claimed to have coupled the NAD derivative to Asp-273 positioned slightly outside the active site. He explained the negative results of Mansson et al. by the blocking of the carboxyl group of Asp-223 in the active center by the pyrazole molecule and the inaccessability of the primary amino group of N^6-[N-(6-aminohexyl)-carbamoylmethyl]-NAD to Asp-273 because of the too short spacer length of the latter NAD analog.

Actually, this example emphasizes the general importance of the spacer length of a functional group for an effective preparation of active enzyme-coenzyme conjugates by ternary complexes. Usually, NAD covalently coupled to the dehydrogenase was not totally reducible by substrate. The percentage of substrate-reducible NAD was found to be 25% for alcohol dehydrogenase [93], 40–60% for lactate dehydrogenase [96], 60–90% for malate dehydrogenase [97], and 70% for glucose dehydrogenase [98]. Therefore, a considerable part of the bound NAD analogs does not interact with the active site, presumably, due to the shortness of the spacers employed. All this may reflect a certain randomness in the attachment of the reactive coenzyme analog, especially if the ternary complex strategy has not been followed. A longer spacer between coenzyme analog and point of attachment on the polypeptide chain of the enzyme may favor the reducibility of enzyme-NAD complexes as seen for malate dehydrogenase and glucose dehydrogenase. Actually, these enzyme-NAD complexes were not prepared using the ternary complex approach.

The strategy of formation of ternary complexes, before the reaction of the covalent attachment of the functionalized NAD is initiated, may lead to even better reducible enzyme-NAD conjugates, if the reactive NAD analogs contain appropriate spacers that facilitate the in-out movement with respect to the active site for the coupled coenzyme.

It should be emphasized that dehydrogenase-NAD(H) complexes can be present in the open (coenzyme dissociated from the active site) and in the closed form (coenzyme associated with the active site). Self-evidently, the tendency to dissociate from the active site is greater for enzyme-coupled NAD than for the NADH form, since the dissociation constants for NAD are generally 10 to 20-fold higher than for NADH.

Steady-state kinetics showed that the ratios open complex/closed complex for horse liver alcohol dehydrogenase-NAD and -NADH complex were around 0.3 and approx. 0 [96]. These complexes were synthesized according to Mansson et al. [92] by using short-spacered reactive NAD. For thermostable malate dehydrogenase-NADH complex where a long spacer (25 nm) was integrated, Eguchi et al. [97] could show that the open form was indeed better represented, giving a ratio of 0.36. This is in the same range as observed for the alcohol dehydrogenase-NAD complex with a short spacer. The occupancy of the active site by the covalently linked coenzyme has been qualitatively demonstrated by other findings such as increased K_M values for exogenous NAD [96,97], reduced inhibition by a competitive inhibitor [92], weaker interaction with affinity ligands [97] and increased thermostability of the enzyme [96].

One of the striking properties of dehydrogenase-NAD complexes is the intramolecular character of the reaction between the NAD moiety and the enzyme moiety, unlike the intermolecular reaction between free NAD and enzyme. Eguchi et al. [97] and Nakamura et al. [98], converting L-malate or glucose in the presence of a chemical redox system for NAD recycling and monitoring the reaction, proved that the rate constants for malate dehydrogenase-PEG (M_r 3,000)-NAD and glucose dehydrogenase-PEG (M_r 3,000)-NAD complexes (spacer 25 nm) were independent of the concentration of the complexes, showing first-order kinetics. In contrast, the results of control experiments with free NAD and enzyme at a fixed ratio pointed to second-order kinetics. For enzyme-coenzyme complexes, the local effective concentration of the coenzyme is recognized as being constant irrespective of the concentration of the enzyme-coenzyme complex. The effective concentration of NAD in dehydrogenase complexes can be estimated from inhibition experiments by studying the effect of the exogenous NAD concentration on the activity of the enzyme-NAD complex by comparison with the native enzyme under conditions where the enzyme coupled coenzyme is in the NADH form and acts as competitive inhibitor. The inhibitor concentration I ($=C_e$, the effective local coenzyme concentration of the enzyme-NAD complex) can be calculated from the expression:

$$[I] = C_e = K_i(K_M'/K_M - 1) \; [97]$$

$K_i \approx K_M^{PEG(M_r\,3,000)-NADH}$ for the native enzyme
$K_M' = K_M^{NAD}$ (apparent) for the enzyme-PEG (M_r 3,000)-NADH complex
$K_M = K_M^{NAD}$ for the native enzyme

For male dehydrogenase-PEG (M_r 3,000)-NAD and glucose dehydrogenase -PEG (M_r 3,000)-NAD complexes effective local concentrations of, respectively, 4×10^{-5} M [97] and 4.2×10^{-3} M [98] have been estimated. It is obvious that for glucose dehydrogenase the reaction rate is greatly enhanced simply by linking a reactive PEG-NAD derivative to the enzyme (anchimeric assistance effect in connection with the intramolecular character of the enzyme-coenzyme interaction). Important is that the enzyme part of the glucose dehydrogenase-PEG (M_r 3,000)-NAD complex recognizes the coupled coenzyme as if the concentration was 4.2×10^{-3} M in solution, although the affinity of the coupled NAD analog for the active site of the enzyme is dramatically decreased because of the N^6 functionalization of the adenine. For uncoupled PEG (M_r 3,000)-N^6-(2-carboxyethyl)-NAD utilized as coenzyme for glucose dehydrogenase extremely high K_M values ($> 4.2 \times 10^{-3}$ M) must be assumed [98]. Nakamura et al. [98] report a 10,000-fold increase of the reaction rate at low glucose dehydrogenase-PEG (M_r 3,000)-NAD concentration (e.g., 0.31×10^{-6} M) compared to concentrations of glucose dehydrogenase (0.31×10^{-6} M) + uncoupled PEG (M_r 3,000)-NAD (0.65×10^{-6} M). This example demonstrates the potential of covalently linking a coenzyme to an enzyme for improving the catalytic effectivity of coenzyme-dependent enzymes.

For eventual applications, an enzyme-coenzyme conjugate has to act as polymer-bound coenzyme with respect to a second enzyme with the same coenzyme requirement. Regeneratable alcohol dehydrogenase-NADH or lactate dehydrogenase-NADH could drive L-lactate formation by lactate dehydrogenase or propanol formation by alcohol dehydrogenase [96]. Malate dehydrogenase-PEG (M_r 3,000)-NADH, autooxidizable by oxaloacetate, could function as regeneratable polymer-bound NAD for the NAD-dependent oxidation of ethanol to acetaldehyde as model reaction catalysed by alcohol dehydrogenase [97]. Under identical conditions, including the coenzyme concentration (enzyme-bound or native), L-malate was continuously produced in batch using malate dehydrogenase-NAD(H) with a rate of around 50% of that for native NAD. Kato et al. [99] studied the formate dehydrogenase-NAD conjugate as exogenous regeneratable polymer-bound NADH, e.g. for the NADH-dependent reductive amination of α-ketoisocaproic acid to leucine by leucine dehydrogenase by comparison with the formate dehydrogenase/leucine dehydrogenase/native NAD(H) system in 10-ml batch experiments under identical conditions, except for the coenzyme concentration. The NAD concentration in the reaction mixture, containing formate dehydrogenase-NAD complex, was around 3.3×10^{-6} M, which was 3.3% of the K_M^{NAD} for formate dehydrogenase and 9.4% of the K_M^{NADH} for leucine dehydrogenase, unlike the conditions for the native enzyme system with a higher NAD concentration of 0.6×10^{-3} M. Unexpectedly, the rate of leucine formation with the formate dehydrogenase-NAD conjugate was much higher than predicted from the low coenzyme concentration, resulting in 15% of the rate for the formate dehydrogenase/leucine dehydrogenase/NAD system with the formate dehydrogenase-NADH concentration as limiting factor for α-ketoisocaproic acid conversion. Presumably, the formate dehydrogenase-NAD conjugate is another example of an enzyme-coenzyme complex that reveals improved catalytic activity by the anchimeric assistance effect.

The coupled substrate approach has been demonstrated for alcohol dehydrogenase-NAD(H) producing propandiol from lactaldehyde [92] or cinnamaldehyde from

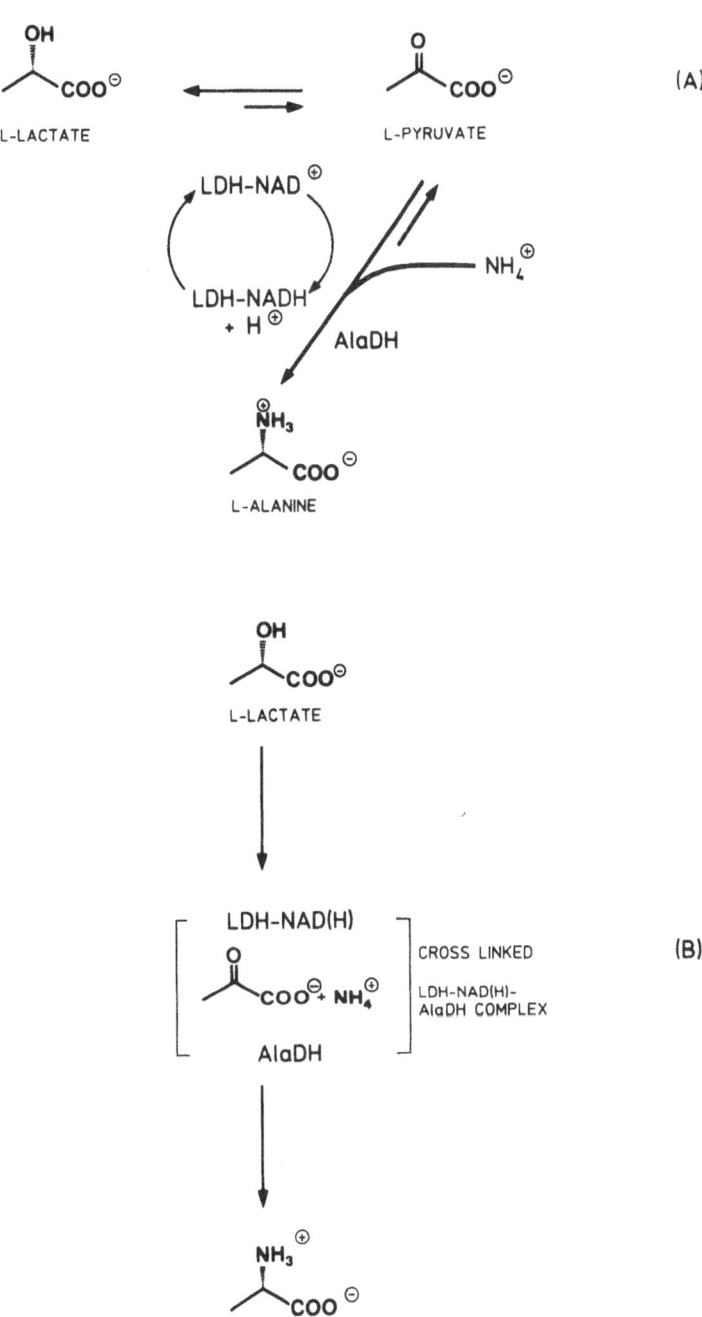

Fig. 9A, B. Pyruvate as substrate for the coupled reactions catalysed by lactate dehydrogenase (LDH)-NAD(H) conjugate and alanine dehydrogenase (Ala DH) producing L-alanine without (A) and with (B) substrate tunneling; K [overall] $= 0.9 \times 10^9$ M^{-1}

cinnamic alcohol [100] by regenerating respectively alcohol dehydrogenase-NADH with ethanol or alcohol dehydrogenase-NAD with acetaldehyde.

Another representative variant is the system where L-lactate is converted to L-alanine with pyruvate as intermediate product/substrate using lactate dehydrogenase-NAD(H) complex in combination with L-alanine dehydrogenase with the equilibrium of the overall reaction on the side of L-alanine (Fig. 9A). The increased thermostability of lactate dehydrogenase-NAD(H) and alanine dehydrogenase suggests a close interaction of the coenzyme with both enzymes [96]. Recycling of the lactate dehydrogenase-bound NAD by alanine dehydrogenase involves the diffusion of large molecules and 50% lower cycling rates [600–800 mol product per mol coenzyme per h] have been observed compared to similar enzyme systems using free NAD(H) [96]. Jacobi and Woenckhaus [102] claim to have increased the cycling rate substantially using site-to-site oriented complexes of chemically crosslinked lactate dehydrogenase-NAD and alanine dehydrogenase. Presumably, this site-to-site orientation enabled a direct back-and-forth transfer of lactate dehydrogenase-coupled NAD(H) between the cooperating enzymes within the crosslinked enzyme complex and substrate tunneling for pyruvate during the reaction cycle, as shown by a very low pyruvate concentration in the reaction medium (Fig. 9b) [102].

The use of enzyme-coenzyme complexes of NADP or FAD in the field of enzymatic synthesis or analysis (biosensors) requires further investigation.

Increasing knowledge of the three-dimensional structure of coenzyme-dependent enzymes, especially with respect to the spatial active site-coenzyme interaction, will surely lead to more effective procedures for the preparation of useful enzyme-coenzyme conjugates with emphasis on the design of the reactive coenzyme analog (spacer length) and the search for the right amino acid residue for achieving appropriate covalent attachment.

The improvement of the intramolecular coenzymatic activity for enzyme-coenzyme complexes by anchimeric assistance may have significant impact on the development of new concepts for the application of coenzyme-dependent enzymes [98].

4 Properties of Water-Soluble Macromolecular Redox Coenzyme Derivatives

In this chapter, those properties of macromolecular coenzyme derivatives will be briefly considered that are of direct interest for application to reactor concepts for coenzyme-dependent enzymes based on ultrafiltration.

4.1 Retention

Reactor concepts based on ultrafiltration pose high demands on the retention of both participating enzymes and macromolecular coenzyme derivatives. Most of the macromolecular coenzyme derivatives described in Chap. 3 will be practically 100%

retained by commercially available ultrafiltration membranes (hollow fiber or flat-bed membrane mode) with cut-offs in the molecular weight range 5,000–10,000 for proteins. The use of inert uncharged membranes for the charged macromolecular coenzyme derivatives applies as a rule of thumb. Bulky polymer-bound coenzyme derivatives with branched chains in the backbone, e.g. polythelene imine, will not be problematic with respect to retention.

Nonbulky macromolecular coenzyme derivatives of moderate molecular weight may imply some risk with respect to retention. Wandrey and Wichmann [103] have studied the elution behavior of polyethylene glycol (PEG) (M_r 10,000)- and PEG (M_r 20,000)-N^6-(2-aminoethyl)-NADH in the absence of protein for an enzyme membrane reactor with an ultrafiltration membrane of low cut-off (5,000). A considerable difference in elution loss was found in favor of PEG (M_r 20,000)-N^6-(2-aminoethyl)-NADH (e.g., respectively 4.3% and 1.7% loss per day with residence time 1 h, corresponding to a retention of 99.82% and 99.93%). Anticipating even better retention under operation conditions in the presence of protein, PEG (M_r 20,000)-N^6-(2-aminoethyl)-NADH was chosen as coenzyme for membrane reactors with continuously operating NAD(H)-dependent enzyme systems.

Concluding, it should be stressed, that the residence time is an important factor for judging the retention properties of water-soluble macromolecular coenzyme derivatives in those cases where a 100% retention is doubtful in advance. Especially, if short residence times are required from the economic point of view (high space time yields), dubious macromolecular coenzymes must be avoided to ensure satisfactory retention.

4.2 Functional and Chemical Stability

The functional and chemical stability of macromolecular coenzyme derivatives have been considered in Sects. 2.1 and 3.1 without stressing their differences.

The functional stability is related to alterations based on nonintrinsic factors, for example lack of selectivity of the coenzyme regeneration method or reaction of the substrate with the coenzyme withdrawing active coenzyme under operating conditions [6].

The chemical stability deals intrinsically with the macromolecular coenzyme derivative itself, e.g. stability of the linkage between polymer and functionalized coenzyme analogue or between the functional group and the adenine part of the coenzyme and of the nicotinamide or isoalloxazine moiety of NAD(P) (H) or FAD in the reduced or oxidized state.

Self-evidently, macromolecular coenzymes ought to be synthesized and applied under favorable conditions with respect to both kinds of stability.

The fact that both NAD and NADH forms are present under steady-state conditions in a continuously producing membrane reactor with NAD(H)-dependent enzyme systems [103], it should be emphasized that macromolecular NAD derivatives do show a remarkably better long-term stability under the usual moderate alkaline conditions (pH 7.5–9) than native NAD as shown by Bückmann [66] and later verified by others [80, 104] (Fig. 10).

REDUCABILITY (%)

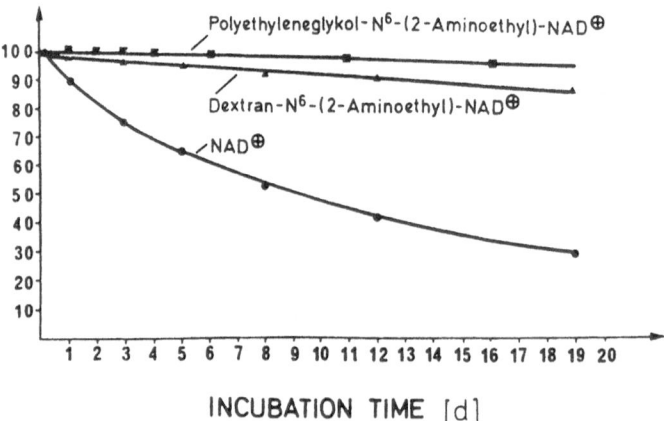

INCUBATION TIME [d]

Fig. 10. Stability of polyethylene glycol (M_r 10,000)-N^6-(2-aminoethyl)-NAD$^+$, dextran (M_r 40,000)-N^6-(2-aminoethyl)-NAD and native NAD$^+$ under moderate alkaline conditions. Incubation conditions: 1 mM NAD$^+$ (polymer-bound or native) in 0.1 M Tris-HCl, pH 9.0, with 0.02% sodium azide at 20 °C. Reducibility test conditions: 0.08 mM NAD$^+$ (polymer-bound or native) in 0.1 M Tris-HCl, pH 8.0, with 0.1 M ethanol, 7 mM semicarbazide, 0.1 mM EDTA, and 2 µM yeast alcohol dehydrogenase at 20 °C

4.3 Coenzymatic Activity

The maintenance of a coenzymatic activity comparable to that of the native coenzyme is a main prerequisite for the application of macromolecular coenzyme derivatives. It is not possible to predict if a coenzyme will have a satisfactory coenzymic activity after modification. Since only the adenine moiety can be involved (Sect. 3.1) in the modification of coenzymes such as NAD(H), NADP(H) and FAD, it is only a question of whether the modified adenine of the coenzyme analogue can still properly interact with the active site. This will totally depend on the enzyme chosen. It must be stressed that knowledge about the coenzyme interaction might be very helpful for such predictions. Up to now, studies with respect to the coenzymatic activity of macromolecular coenzyme derivatives have used trial-and-error approaches with little apparent rationale.

The coenzyme activity for macromolecular coenzyme analogs have been expressed by experimentally obtained K_M and V_{max} values. They are often of apparent character if, for example, part of the polymer-bound coenzyme derivatives has no access to the active site of the enzyme (bulky macromolecular coenzyme derivatives). After macromolecularization of NAD(H) or NADP(H), several effects on the coenzyme activity have been demonstrated, ranging from negligible to even improved activity compared to the native coenzymes.

Isocitrate dehydrogenase (NADP-dependent from yeast), 12α-hydroxysteroid dehydrogenase (NADP-dependent from *Clostridium* P, C 48–50), and glucose dehydrogenase (NAD(P)-dependent from *Bacillus megatherium*) show no or extremely low activity with PEG (M_r 3,000)-N^6-(2-carboxyethyl)-NADP [60] or PEG (M_r

20,000)-N^6-(2-aminoethyl)-NADP [42], but also with N^6-(2-carboxyethyl)-NADP [60] or N^6-(2-aminoethyl)-NADP [42], in the usual coenzyme concentration range 0.05–1 mM. For glucose dehydrogenase, PEG (M_r 20,000)-N^6-(2-aminoethyl)-NAD and N^6-(2-aminoethyl)-NAD were somewhat better coenzymes since high K_M value were measured (respectively, 2 and 4.5 mM versus $K_M^{NAD} = 0.064$ mM) [42].

These results suggest that intrinsically the N^6-modification of the adenine has dramatically changed the coenzyme activity by weakening the glucose dehydrogenase coenzyme interaction. Weak but still acceptable activity has been observed for alanine dehydrogenase from *Bacillus subtilis* using PEG (10,000)-N^6-(2-aminoethyl)-NADH as coenzyme (Fig. 11) [81]. Almost equal coenzyme activity has been observed for leucine dehydrogenase from *B. sphaericus* with PEG (M_r 10,000)-N^6-(2-aminoethyl)-NADH as coenzyme (Fig. 12) [81]. Surprisingly, PEG (M_r 10,000)-N^6-(2-aminoethyl)-NAD turned out to be an even better coenzyme than NAD with respect to V_{max} for formate dehydrogenase from *Candida boidinii* (Fig. 13) [81]. This was not a single phenomenon, as Katayama et al. [36] showed that PEG (M_r 3,000)-N^6-(2-carboxyl-ethyl)-NAD was also a better coenzyme than NAD for malate dehydrogenase from *Thermus thermophilus* and for glyceraldehyde-3-phosphate dehydrogenase from *Bacillus stearothermophilus* (V_{max}/V_{max}^{NAD} approx. 2, K_M values in the same range as K_M^{NAD} for both dehydrogenases).

The result that the coenzyme activity of PEG (M_r 3,000)-N^6-(2-carboxyethyl)-

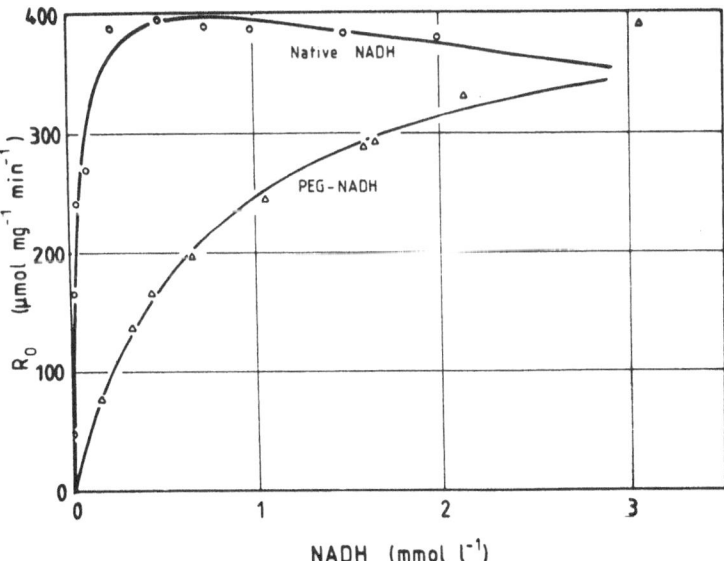

Fig. 11. Michaelis-Menten plot for L-alanine dehydrogenase (*Bacillus subtilis*) with variable concentration of NADH or polyethylene glycol (M_r 10,000)-N^6-(2-aminoethyl)-NADH (range 0–2.5 mM) at fixed concentrations of pyruvate and ammonia. The assay buffer contained 33.3 mM sodium phosphate, pH 8.0, 10 mM sodium pyruvate, 0.4 M NH$_4$Cl, and 0.25 μg L-alanine dehydrogenase per ml. Temperature: 40 °C. NADH: $V_{max} = 448$ μmol, mg^{-1}, min^{-1} · $K_m = 0.0486$ mM. Polyethylene glycol (M_r 10,000)-N^6-(2-aminoethyl)-NADH: $V_{max} = 433$ μmol mg^{-1} min^{-1}. $K_m = 0.75$ mM

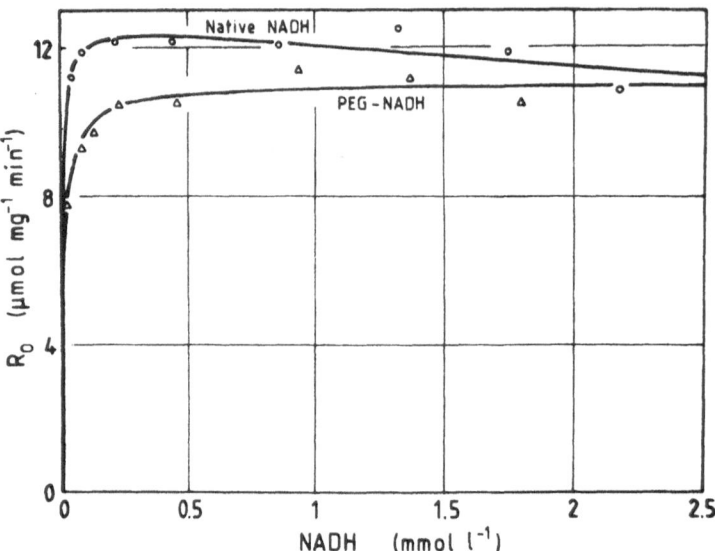

Fig. 12. Michaelis-Menten plot for L-leucine dehydrogenase (*Bacillus sphaericus*) with variable concentration of NADH or polyethylene gylcol (M_r 10,000)-N^6-(2-aminoethyl)-NADH (range 0–2.5 mM) at fixed concentration of α-ketoleucine and ammonia. The assay buffer contained 50 mM sodium phosphate, pH 8.0, 4.95 mM α-ketoleucine, 0.4 M NH_4Cl, and 2 µg L-leucine dehydrogenase per ml. Temperature: 25 °C. NADH: V_{max} = 11.6 µmol mg^{-1} min^{-1}. K_m = 0.0063 mM. Polyethylene glycol (M_r 10,000)-N^6-(2-aminoethyl)-NADH: V_{max} = 11.0 µmol mg^{-1} mg^{-1} min^{-1}. K_m = 0.0134 mM

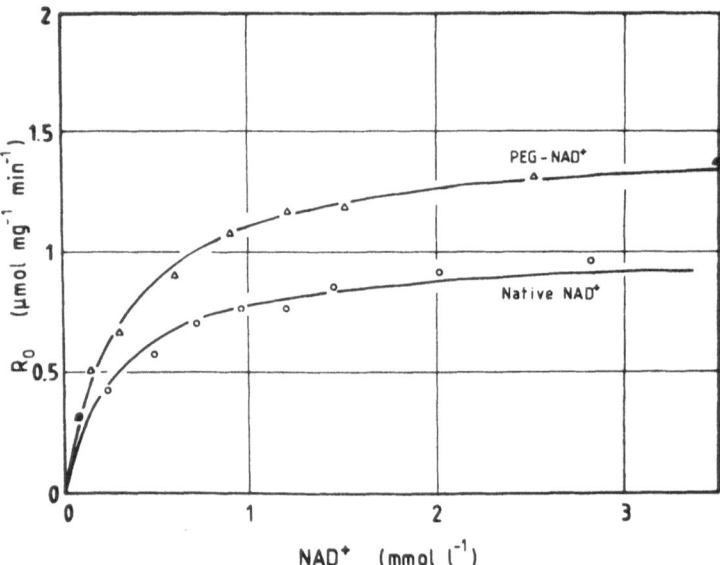

Fig. 13. Michaelis-Menten plot for formate dehydrogenase (*Candida boidinii*) with variable concentration of NAD^+ or polyethylene glycol (M_r 10,000)-N^6-(2-aminoethyl)-NAD^+ (range 0–3.5 mM) at fixed formate concentration. The assay buffer contained 33.3 mM sodium phosphate, pH 8.0, 0.4 M sodium formate, and 2 µg formate dehydrogenase per ml. Temperature: 40 °C. NAD^+: V_{max} = 1.01 µmol mg^{-1} min^{-1}. K_m = 0.296 mM. Polyethylene glycol (M_r 10,000)-N^6-(2-aminoethyl)-NAD^+: V_{max} = 1.48 µmol mg^{-1} min^{-1}. K_m = 0.333 mM

NADP was 40 times that of NADP and 14% that of NAD for NAD specific alcohol dehydrogenase from yeast [36] is a striking example, illustrating that unexpected activity phenomena are possible for macromolecular coenzyme derivatives.

The coenzymatic activity of most of the functionalized NAD(H) and corresponding water-soluble macromolecular derivatives synthesized so far are shown in Table 3 where they are compared with the native coenzymes.

It can be seen that the enzymes, most frequently utilized for testing the activity of coenzyme derivatives, are alcohol dehydrogenase (from yeast and horse liver), lactate dehydrogenase, malate dehydrogenase, and glutamate dehydrogenase. In Table 3, the results are grouped in such a way that the activity of each enzyme with the various NAD(H) derivatives can easily be compared.

The factors that can influence the coenzymatic properties (K_m and V_{max}) of NAD analogs are mainly the position of functionalization (N-1, N^6 or C-8 of the adenine ring), the charge, hydrophobicity or hydrophilicity of the spacer arm and the nature of the polymer. Other factors to be considered are the molecular weight and the NAD density (mol NAD per mol monomer) of the polymeric derivatives. Concerning the position of functionalization, N-1 derivatives (not shown in Table 3) are very poor substrates for NAD-dependent enzymes [75,105], probably because of their inability to form a hydrogen bond between the N-1 atom of the adenine moiety and the appropriate part of the active site of NAD-dependent dehydrogenases [106,107].

For the other two functionalizations, the N^6 position appears preferable to the C-8 position which gives generally lower activity and much higher K_m values, particularly, when linked to polymers [75].

The effects of the charge (neutral, negative or positive), hydrophobicity and length of the spacer arms were investigated in some detail by several authors [57,80,105,108] with macromolecular analogues of NAD substituted at the N^6 position. However, no general rule can be derived, since the response is highly dependent on the enzymes utilized.

Looking at the data summarized in Table 3, a trend becomes visible. In most cases, a decrease of V_{max} is accompanied by an increase of K_M.

Among the soluble polymers [polyethylene glycol (PEG), polyethyleneimine (PEI), dextran] employed to prepare macromolecular NAD derivatives, PEG linked to the N^6 position leads to coenzyme derivatives that give the best results for most of the enzymes listed in Table 3. It is shown that substantial activity is maintained with moderate increase of K_m values. PEI-NAD derivatives give more erratic results, since they posses very good coenzymatic properties with some enzymes (alcohol dehydrogenase from yeast, lactate dehydrogenase from rabbit muscle and formate dehydrogenase from yeast), whereas they are only weakly active with others (glutamate dehydrogenase, alanine dehydrogenase and 3α-hydroxysteroid dehydrogenase). Dextran-NAD derivatives were mainly studied by Schmidt and Grenner [64] who found for the six enzymes examined an activity that was 8.5–37% of that of unmodified NAD (Table 3). Macromolecular coenzymes, synthesized by copolymerization of acrylic monomers with NAD(H) functionalized at the N^6 position have been tested with several enzymes and gave activities that were 5–62% of those found with unmodified NAD with a marked increase of K_m values in the majority of cases (Table 3).

The effect of the molecular weight of the polymer on the coenzymatic properties

Table 3. Coenzymatic properties of functionalized and macromolecular NAD(H) derivatives with respect to some current NAD(H)-dependent dehydrogenases

Enzyme	NAD(H) derivative	K_m^a (μM)	V^b	Refs.
Alcohol DH (horse liver)	N^6-carboxymethyl-NAD		0.55	57)
	N^6-[N-(6-aminohexyl)-acetamide]-NAD		1.00	57)
	N^6-(2-aminoethyl)-NAD	250(180)	0.58	105)
	Dextran-N^6-(2-aminoethyl)-NAD		0.17	64)
	N^6-(2-carboxyethyl)-NAD		0.51	111)
	PEG-N^6-(2-carboxyethyl)-NAD	19(18)	0.66	82)
	Copolymer of methacrylyl choline and N^6-NADH	21(10)	0.28	89)
Alcohol DH (yeast)	C(8)-Br-NAD	170(55)	1.00	71)
	N^6-(2-hydroxy-3-carboxypropyl)-NAD		0.65	72)
	PEI-N^6-(2-hydroxy-3-carboxypropyl)-NAD		1.00	72)
	C(8)-(2-carboxyethylthio)-NAD		0.41	72)
	PEI-C(8)-(2-carboxyethylthio)-NAD		0.47	72)
	N^6-succinyl-NAD	58(92)	0.67	70)
	PEI-N^6-succinyl-NAD	53(92)	0.97	70)
	N^6-(2-aminoethyl)-NAD	330(170)	0.23	105)
	Dextran-N^6-(2-aminoethyl)-NAD		0.08	64)
	N^6-(2-carboxyethyl)-NAD		0.64	111)
	PEG-N^6-(2-carboxyethyl)-NAD		0.50	111)
	Copolymer of methacrylyl choline and N^6-NADH	470(82)	0.37	89)
	Homopolymer of N^6-(acryloxy-2-hydroxypropyl)-NAD		0.81	88)
	N^6-[N-[2-[N-(2-methacrylamidoethyl) carbamoyl] ethyl] carbamoylmethyl]-NAD	500(290)	0.57	108)
	Copolymer of acrylamide with N^6-[N-[2-[N-(2-methacrylamidoethyl) carbamoyl] ethyl] carbamoylmethyl]-NAD	1400(290)	0.28	108)
Lactate DH (beef heart)	N^6-carboxymethyl-NAD		0.65	57)
	N^6-[N(6-aminohexyl)-acetamide]-NAD		0.50	57)
Lactate DH (chicken M_4)	C(8)-Br-NADH	18(10)	1.00	71)
Lactate DH (rabbit muscle)	N^6-(2-hydroxy-3-carboxypropyl)-NAD		0.83	72)
	PEI-N^6-(2-hydroxy-3-carboxypropyl)-NAD		0.60	72)
	C(8)-(2-carboxyethylthio)-NAD		0.58	72)
	PEI-C(8)-(2-carboxyethylthio)-NAD		0.03	72)
	N^6-(2-aminoethyl)-NAD	370(90)	0.28	105)
	Dextran-N^6-(2-aminoethyl)-NAD		0.24	64)
	N^6-(2-carboxyethyl)-NAD		0.77	111)
	PEG-N^6-(2-carboxyethyl)-NAD	51(91)	0.27	82)
	Copolymer of methacrylyl choline and N^6-NADH	45(3)	0.05	89)
	N^6-[N-[2-[N-(2-methacrylamidoethyl) carbamoyl] ethyl] carbamoylmethyl]-NAD	160(250)	0.64	108)
	Copolymer of acrylamide with N^6-[N-[2-[N-(2-methacrylamidoethyl) carbamoyl] ethyl] carbamoylmethyl]-NAD	1200(250)	0.28	108)
Malate DH (pig heart)	N^6-carboxymethyl-NAD		0.75	57)
	N^6-[N(6-aminohexyl)acetamide]-NAD		0.75	57)
	N^6-(2-aminoethyl-NAD	80(75)	0.58	105)
	Dextran-N-(2-aminoethyl)-NAD		0.18	64)

Table 3. (continued)

Enzyme	NAD(H) derivative	K_m^a (µM)	V^b	Refs.
	N^6-(2-carboxyethyl)-NAD		0.62	111)
	PEG-N^6-(2-carboxyethyl)-NAD		0.64	111)
	Copolymer of methycrylyl choline and N^6-NADH	13(13)	0.26	89)
	N^6-[N-[2-[N-(2-methacrylamidoethyl) carbamoyl] ethyl] carbamoylmethyl]-NAD	400(170)	0.91	108)
	Copolymer of acrylamide with N^6-[N-[2-[N-(2-methacrylamidoethyl) carbamoyl] ethyl] carbamoylmethyl]-NAD	1200(170)	0.62	108)
Glutamate DH	N^6-(2-aminoethyl)-NAD	50(160)	0.40	105)
(beef liver)	Dextran-N^6-(2-aminoethyl)-NAD		0.37	64)
	PEG-N^6-(2-aminoethyl)-NAD	719(203)	0.39	75)
	PEI-N^6-carboxymethyl-NAD	75(203)	0.04	75)
	PEG-C(8)-(6-aminohexyl)-amino-NAD	3190(203)	0.69	75)
	Copolymer of methyacrylyl choline and N^6-NADH	79(28)	0.39	89)
Alanine DH	N^6-(2-hydroxy-3-carboxypropyl)-NAD		0.32	61)
(B. subtilis)	PEI-N^6-(2-hydroxy-3-carboxypropyl)-NAD		0.02	61)
	PEG-N^6-(2-aminoethyl)-NADH	750(49)	0.96	81)
	Homopolymer of N^6-(acryloxy-2-hydroxypropyl)-NAD		0.45	88)
			0.45	88)
Formate DH (C. boidinii)	PEG-N^6-(2-aminoethyl)-NAD	333(296)	1.46	81)
Formate DH	PEG-N^6-(2-aminoethyl)-NAD	82(15)	0.57	75)
(yeast)	PEI-N^6-carboxymethyl-NAD	5(15)	0.85	75)
	PEG-C(8)-(6-aminohexyl)-amino-NAD	2450(15)	0.03	75)
3α-hydroxy-	PEG-N^6-(2-aminoethyl)-NAD	647(38)	0.66	75)
steroid DH	PEI-N^6-carboxymethyl-NAD	108(38)	0.08	75)
(P. testosteroni)	PEG-C(8)-(6-aminohexyl)-amino-NAD	6300(38)	0.51	75)
Glyceraldehyde	N^6-[N-(N-acryloyl-1-methoxycarbonyl-5-aminopentyl)-propionamide]-NAD		0.32	59)
-3-phosphate DH				
(rabbit muscle)	Copolymer of acrylamide with N^6-[N-(N-acryloyl-1-methoxycarbonyl-5-aminopentyl)-propionamide]-NAD		0.06	59)
Leucine DH (B. sphaericus)	PEG-N^6-(2-aminoethyl)-NADH	13(6)	0.95	81)
Aldehyde DH (yeast)	C(8)-Br-NAD	440(55)	0.80	71)

a In parentheses K_m values for unmodified NAD(H);
b rate relative to unmodified NAD(H);
DH: Dehydrogenase;
PEG: Polyethylene glycol;
PEI: Polyethyleneimine

was investigated by Schmidt and Dolabdjian [109] and Adachi et al. [104] with dextran-NAD derivatives of molecular weight 8,500–600,000 using yeast alcohol dehydrogenase. The results are contradictory, since the first authors [109] found that the coenzymatic activity decreased on increasing the molecular weight, whereas Adachi et al. [104] found that the molecular weight had no influence. Contradictory results

Table 4. Coenzymatic properties of functionalized and macromolecular NADP derivatives with respect to some current NADP-dependent dehydrogenases

NACP derivative	Glucose-6-phosphate DH (yeast)		Glutamate DH (beef liver)		Aldehyde DH (yeast)		6-Phospho-gluconate DH (yeast)	Alcohol DH (L. Mesenteroides)	Refs.
	K_m^a (µM)	V^b	K_m (µM)	V	K_m (µM)	V	V	V	
N(1)-carboxymethyl-NADP		1.45		0.50			0.60		58)
N⁶-carboxymethyl-NADP		0.65		0.35			0.80		
C(8)-(6-aminohexyl)-amino-NADP	44(36)	0.42							71)
N(1)-(2-hydroxy-3-carboxypropyl)-NADP		1.00		0.17		0.67			62)
N⁶-(2-hydroxy-3-carboxypropyl)-NADP	12(14)	0.80	395(164)	0.69	126(53)	0.53			62)
C(8)-(2-carboxyethylthio)-NADP	12(14)	0.48	391(164)	0.84	157(53)	0.33			62)
PEI-N(1)-(2-hydroxy-3-carboxypropyl)-NADP		0.22		0.23		0.36			62)
PEI-N⁶-(2-hydroxy-3-carboxypropyl)-NADP		0.16		0.03		0.29			62)
PEI-C(8)-(2-carboxyethylthio)-NADP		0.10		0.02		0.01			62)
N⁶-(2-carboxyethyl)-NADP	3(9)	0.72	18(5)	1.25			0.80	0.58	74)
PEG-N⁶-(2-carboxyethyl)-NADP	10(9)	0.89	21(5)	0.82		0.16	1.05	0.53	60)

^{a} In parentheses K_m values for unmodified NADP;
^{b} rate relative to unmodified NADP;
DH: Dehydrogenase;
PEG: Polyethylene glycol;
PEI: Polyethyleneimine

were also reported as to NAD density. Yamazaki and Maeda [87] and Furukawa et al. [110,135] found that the coenzymatic activity decreased with an increase in the NAD density (4×10^{-4} to 6.5×10^{-2} range) in the case of copolymerized spacered NAD (M_r 100,000), e.g. for yeast alcohol dehydrogenase and pig heart L-lactate dehydrogenase. On the contrary, Adachi et al. [104] observed an increased and constant coenzymatic activity (approx. that of NAD), respectively, in the density ranges 1×10^{-2} to 1×10^{-1} and 1×10^{-1} to 2.5×10^{-1} for the same enzymes, using dextran (M_r 10,000–70,000)-NAD with a spacer of similar length compared to copolymerized NAD.

Katayama et al. [36] also found that the source of the enzyme can influence its response with macromolecular NAD(H) analogues. In fact, alcohol dehydrogenase, lactate dehydrogenase, malate dehydrogenase, and glyceraldehyde-3-phosphate dehydrogenase from different sources (prokaryotic and eukaryotic, mesophilic and thermophilic organisms) gave rather different results and two of them (malate dehydrogenase form *Thermus thermophilus* and glyceraldehyde-3-phosphate dehydrogenase from *B. stearothermophilus*) showed an activity with NAD analogues even higher than with unmodified NAD.

With regard to macromolecular NADP, fewer derivatives have been synthesized and fewer enzymes have been tested compared to macromolecular NAD (Table 4). Unlike the corresponding NAD analogues, NADP analogues functionalized at the N(1) position of the adenine moiety are rather good coenzymes. Of the two polymers PEG and PEI used to enlarge the molecular-weight of the coenzyme, PEG gives better results. In fact, PEG-N^6-(2-carboxyethyl)-NADP shows high activity (except for aldehyde dehydrogenase) with a slight increase of K_m values compared to unmodified NADP (Table 4) [60].

Up to now, only Zappelli et al. [63] have reported on the function of macromolecular N^6-modified FAD as coenzyme for oxidases with noncovalently bound FAD. As an example, PEI (M_r 40,000–60,000)-N^6-(2-hydroxy-3-carboxypropyl)-FAD entrapped together with D-amino acid oxidase ($K_d^{FAD} = 0.5 \times 10^{-6}$ M) in triacetate fibers could considerably prolong the operational stability of both the apo- and holoenzyme, functioning, respectively, directly as coenzyme and as substituent for slowly bleeding-out FAD. Furthermore, it was found that PEI-bound N^6-(2-hydroxy-3-carboxypropyl)-FAD could better reactivate the apoenzyme of glucose oxidase and D-amino oxidase than PEI-bound N(1)-(2-hydroxy-3-carboxypropyl)-FAD (respectively, 97 % and 20 % versus 15 % and 5 % reactivation), indicating a similar preference pattern with respect to interaction as was observed on comparing macromolecular N^6- and N(1)-functionalized NAD derivatives.

Concerning the coenzymatic activity of macromolecular coenzyme derivatives, the following rule of thumb may apply:

If a low-molecular-weight functionalized coenzyme derivative shows satisfactory coenzyme activity with a certain enzyme, it may be anticipated that a corresponding not too bulky macromolecular coenzyme derivative will also show similar coenzymatic activity, as expressed by K_M and V_{max}.

Factors such as charge, hydrophobicity or hydrophilicity of the spacer between polymer backbone and coenzyme analogue, molecular weight of the polymer and coenzyme density, are of secondary importance. They should, if necessary, be separately studied for each enzyme.

5 Application of Water-Soluble Macromolecular Redox Coenzyme Derivatives

5.1 General Aspects

The cardinal question with respect to the application of coenzyme-dependent enzymes with simultaneous enzymatic coenzyme regeneration is the choice of the reactor concept. Preferably, a concept for continuous operation should be chosen. In fact, a choice must be made between utilizing whole cells with naturally present coenzyme-dependent enzyme systems or combining appropriate reactions outside the cell environment to be catalysed by isolated and partially purified enzymes which then may come from different sources. In order to evaluate the advantages and disadvantages of utilizing whole cells versus isolated enzyme systems for coenzyme-dependent biotransformations, the informative table given by Wandrey and Wichmann [103] may be helpful.

If the downstream processing of the product is not complicated by compounds arising from secondary reactions and an acceptable space-time yield can be achieved, the utilization of whole cells should be preferred, since the coenzyme regeneration is solved intracellularly.

For whole cells, the reactor concept may be based on current bioprocess technology for suspended cells in solution (semi-continuous or continuous mode) or on immobilized cells (viable or resting cells, permeabilized or not) obtainable by a variety of entrapment or adsorption techniques, mostly leading to continuously operating column reactors.

High expression of appropriate coenzyme-dependent enzymes with high stability and efficient downstream processing procedures for their isolation and purification may be realistic arguments to choose the application of isolated technical enzyme preparations in spite of complicating coenzyme regeneration [41,112].

Isolated and assembled coenzyme-dependent enzymes have been employed in batch to demonstrate the functioning of coenzyme-dependent enzymatic synthesis with simultaneous enzymatic coenzyme regeneration [10,113], and also for the production of various valuable compounds on a mmol scale [20,114]. The main disadvantages of this reactor concept are the eventual product inhibition due to product accumulation, the lack of continuous operation and the difficulty of product separation leaving the enzyme-coenzyme system intact.

Coimmobilization of coenzyme-dependent enzymes for synthesis and simultaneous coenzyme regeneration on suitable carrier material leads to heterogeneous systems that can be run continuously in column form, allowing easy separation of the products from the intact enzymes. As demonstrated by Carrea et al. [115], NADP was used for the synthesis of 12-keto chenodeoxycholic acid by immobilized 12α-hydroxysteroid dehydrogenase and simultaneously regenerated by coimmobilized glutamate dehydrogenase, achieving a TTN of 1.600 for the coenzyme.

The disadvantage of this approach is the need to recover the coenzyme from the product solution by laborious isolation techniques (adsorption, ionexchange) which may be the bottleneck for an efficient use of the coenzyme. An unfavorable factor for good reactor performance, generally inherent in immobilized enzyme systems, may be further the hindered access of substrate and coenzyme molecules to the carrier-fixed enzymes due to steric hindrance and/or limited diffusion.

Alternative methods for the coimmobilization of coenzyme-dependent enzymes and their application have been reported by Kajiwara and Maeda [116].

The droplet entrapment method was introduced to prepare enzyme gel particles with scattered small droplets containing homogeneously dissolved enzymes and NAD. Better retention of enzymes was achieved as with other gel entrapment techniques [117], but the retention problem for the native coenzyme could not be satisfactorily solved. One may wonder if the coentrapment of a suitable macromolecular NAD derivative might have resulted in droplet gels of better applicability for continuous coenzyme-dependent transformations with simultaneous coenzyme regeneration. Actually, Chang [118] took this step by preparing semipermeable microcapsules by interfacial polymerization containing an internal aqueous solution with NAD-dependent enzymes and dextran-NAD surrounded by an ultrathin semipermeable membrane. The application of semipermeable microcapsules as artificial cells for medical purposes will be discussed in Sect. 5.2. The system developed by Chang represents an elegant reactor concept for the application of enzymes based on ultrafiltration.

The enzyme membrane reactor concept [103,119] has proved to be a useful alternative for applying coenzyme-dependent enzyme systems with simultaneous enzymatic coenzyme regeneration in continuous mode. Enzymes and coenzymes present in homogeneous solution are confined within the reaction space of an ultrafiltration reactor system (hollow-fiber or flat-bed membrane configuration) closed off by an appropriate semipermeable ultrafiltration membrane giving a required retention of practically 100%. Substrate solution is continuously pumped into the reactor space, while product solution is leaving at the same flow rate after substrate transformation.

Since the problem of attaining 100% retention of native coenzymes within an ultrafiltration system, while keeping unhindered substrate and product transport, has not been satisfactorily solved, coenzymically active macromolecular coenzyme derivatives are required for the enzyme membrane reactor concept. This concept has the advantage of more or less unhindered interaction between enzymes, substrates and macromolecular coenzyme derivatives due to the homogeneous character of the solution within the reactor, combined with an easy separation of the product from an intact enzyme-coenzyme system.

In this context, a potential complication inherent to the ultrafiltration of proteins should be pointed out. The phenomenon of concentration polarization as a consequence of a one-direction flow of liquid towards the membrane may lead to an increased concentration of enzyme protein and macromolecular coenzyme derivatives in front of the membrane. Especially if the solubility limit is reached and a filter cake is formed (secondary membrane), a mass transport problem may occur for the flux of low-molecular-weight compounds through the membrane. Also the interaction between enzymes, substrates and macromolecular coenzyme derivatives will be disturbed. By circulating the enclosed reactor solution superimposed on the substrate-product flow direction, the negative effect of concentration near the ultrafiltration membrane on the reactor performance can be satisfactorily overcome (Fig. 14) [103].

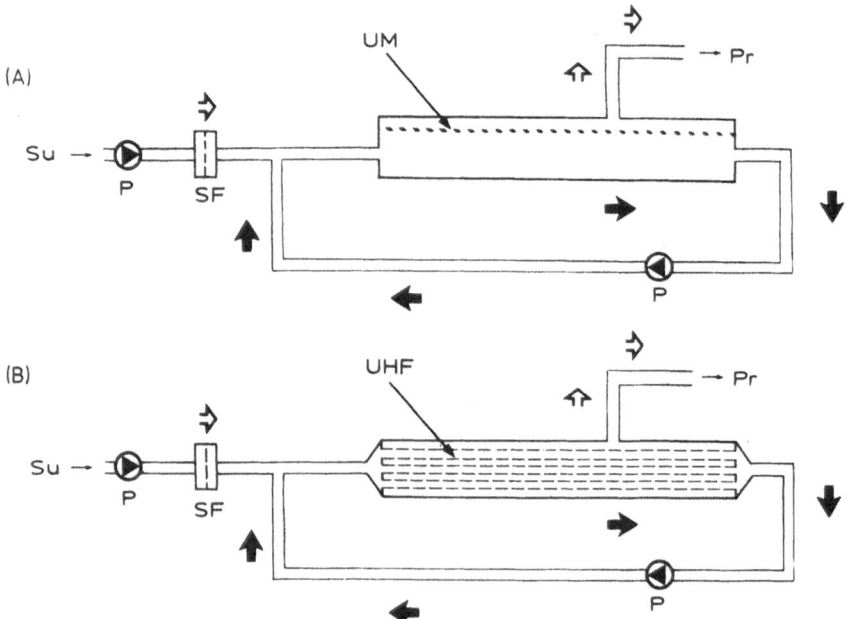

Fig. 14A, B. Schematic presentation of an enzyme membrane reactor: flat-bed membrane mode (A), hollow fibre mode (B); substrate-product flow (⇒); circulating solution with enzymes and, if applicable, macromolecular coenzyme derivatives with flow direction superimposed on the substrate-product flow near the ultrafiltration membrane in the reactor volume (➔)
Su: substrate; Pr: product; SF: sterile filter; P: pump; UM: flat ultrafiltration membrane; UHF: ultrafiltration hollow fibers

The strategy of using isolated coenzyme-dependent enzymes for both the synthetic and the coenzyme regenerating reaction in homogeneous aqueous solution in continuous mode based on the enzyme membrane reactor concept, implies choosing a relatively simple reaction system compared to heterogeneous systems or the complex reaction machinery of a whole cell. The simplicity makes this system describable and, consequently, optimizable by an elementary chemical engineering approach.

During the course of a coenzyme-dependent biocatalytic process with simultaneous enzymatic coenzyme regeneration, the activity of the enzyme catalyzing product synthesis must be matched by the coenzyme-regenerating activity to keep the coenzyme concentration at a catalytic level, while maintaining a satisfactory substrate conversion percentage. Wichmann et al. [13], Fiolitakis and Wandrey [120] and Wandrey and Wichmann [103] describe a chemical engineering approach for the prediction of the optimal performance of an enzyme membrane reactor for the production of L-alanine from pyruvate and L-leucine from α-ketoisocaproate via NADH-dependent reductive amination by alanine dehydrogenase and leucine dehydrogenase, respectively, utilizing PEG (M_r 10,000 and 20,000)-N^6-(2-amino-ethyl)-NADH as coenzyme and formate dehydrogenase for simultaneous NADH regeneration. The enzyme membrane reactor is treated as a continuous-stirred tank reactor (CSTR). By determining the kinetics for the participating enzymes a complete set of K_M and V_{max} values, including non-negligible inhibition constants, with respect to substrates, products and PEG-N^6-(2-aminoethyl)-NAD(H) were obtained. The differential equations for the mass balances for all the convertible components of the system were solved and a reactor model involving the mean residence time was set up to describe and predict the performance of the enzyme membrane reactor under both start-up and steady-state conditions, accounting for the influences of enzyme inactivation and coenzyme loss due to instability and minor passage through the ultrafiltration membrane.

The model was applied to find an optimal reactor size, to predict the optimum ratio for the enzymes, the optimal PEG-N^6-(2-aminoethyl)-NAD(H) concentration and the most favorable substrate concentration in the feed for optimal conversion of the corresponding α-keto acids in L-alanine und L-leucine, and, thus, to design an economically feasible process with optimal space-time yields.

Later this chemical engineering approach was extended to more complicated multienzyme systems, e.g. L-leucine formation from DL-hydroxy isocaproate or L-methionine from DL-α-hydroxy methylthiobutyrate by the D- and L-hydroxy isocaproate dehydrogenase/leucine dehydrogenase/PEG-(M_r 20,000)-N^6-(2-amino-ethyl)-NAD(H) system [121,122] and L-phenylalanine from DL-phenyllactate with the D- and L-hydroxy isocaproate dehydrogenase/L-phenylalanine dehydrogenase/PEG (M_r 20,000)-N^6-(2-aminoethyl)-NAD(H) system [123] with reasonable agreement between experiment and model.

This chemical engineering approach was also successfully utilized (Sect. 5.2) for finding favorable production conditions for the unnatural amino acid L-tert-leucine from 3,3-dimethyl-2-ketobutyric acid on pilot scale (kg quantities) by the leucine dehydrogenase/formate dehydrogenase/technical grade PEG (M_r 20,000)-N^6-(2-aminoethyl)-NAD(H) system. For the model enzyme-coenzyme system L-lactate dehydrogenase/alcohol dehydrogenase/PEG (M_r 3,000)-N^6-(2-carboxyethyl)-NAD(H), producing L-lactate from pyruvate and regenerating PEG-NADH by oxidation of ethanol forming acetaldehyde, Katayama et al. [82] report a similar approach for modelling the steady-state behavior of a continuously operating enzyme membrane reactor. Again agreement was found between experimental results and prediction from the model.

Surprisingly, the merits of enzyme-coenzyme conjugates with respect to the enzyme membrane reactor concept have not been investigated yet. Any minor loss of macro-molecular coenzyme derivative through the ultrafiltration membrane [e.g. 1.7% d^{-1} for PEG (M_r 20,000)-N^6-(2-aminoethyl)-NAD(H) (Sect. 4.1)] may in principle be prevented by applying these particular macromolecular coenzyme derivatives. Maintaining a coenzyme concentration range of $10^{-4} - 10^{-3}$ M for an operating enzyme membrane reactor would mean a high protein content in the reactor solution for the enzyme-coenzyme complex (e.g. M_r 100,000:10–100 mg ml^{-1}) that should also function satisfactorily as a normal polymer-bound coenzyme with respect to a second enzyme in case of coupled two-enzyme systems. The phenomenon of concentration polarization might become insuperably prominent, leading to a break-down of the reactor. Thus far, few results have been published regarding the prospects of applying enzyme-coenzyme complexes in an enzyme membrane reactor. The interaction between an enzyme-coenzyme conjugate and another coenzyme-dependent enzyme does not need to be the bottleneck according to the encouraging findings of Eguchi et al. [97] for the (co)enzymatically active malate dehydrogenase-PEG (M_r 3,000)-NAD conjugate. These show acceptable interaction with alcohol dehydrogenase as reducible and regeneratable macromolecular NAD derivatives due to the long PEG-spacer between enzyme and NAD. For coupled two-enzyme systems, a second enzyme with high specific activity may be desirable. The prospect of applying only an alcohol dehydrogenase-NAD(P) conjugate for biocatalytic transformations with simultaneous coenzyme regeneration according to the coupled substrate approach (Sect. 2.3) may be interesting in the study of ultrafiltration reactor concepts such as enzyme membrane reactors.

The presence of the effect of anchimeric assistance for the enzyme-coenzyme conjugate, as has been discovered for the glucose dehydrogenase-PEG (M_r 3,000)-NAD adduct (Sect. 3.3), will be of general importance whatever enzyme system is chosen. One may speculate that enhancement of the catalytic activity by anchimeric assistance and unhindered interaction with a second enzyme of high specific activity will be favorable prerequisites for the application of enzyme-coenzyme conjugates in ultrafiltration reactors.

The preliminary findings of Kato et al. [99] concerning the reaction capacity of the leucine dehydrogenase/formate dehydrogenase-NAD conjugate system supports this speculation (Sect. 3.4).

5.2 Technical Applications

In spite of the development of many synthetic methods for the preparation of macromolecular coenzyme derivatives, these coenzyme derivatives have been mainly applied to demonstrate the usefulness of ultrafiltration membrane reactors in continuous enzymatic synthesis. Various coenzyme-dependent enzyme systems with simultaneous enzymatic coenzyme regeneration have been investigated on a small scale.

It is not surprising that NAD(H)-dependent enzyme systems have been mostly employed, as NADP(H)-dependent reactions are often inefficient for the production of special chemical compounds and for integration into monitoring devices for analytical use (biosensors).

The lack of accessible NADP(H)-dependent enzymes, catalyzing reactions of potential interest, the high cost of NADP(H) compared to NAD(H) and the more difficult preparation of macromolecular NADP(H) derivatives (Sect. 3.2) has discouraged work on the application of such systems.

Chronologically, Davies and Mosbach [78] and Lowe und Mosbach [58] were the first to demonstrate the applicability of coenzyme-dependent enzymatic conversions with simultaneous enzymatic coenzyme regeneration using macromolecular dextran (M_r 40,000)-NAD or -NADP. In the presence of excess pyruvate or glutamate, glutamate (range 1×10^{-4} to 1×10^{-3} M) and pyruvate (range 2×10^{-5} to 8×10^{-4} M) could be determined with an enzyme electrode. Such electrodes are constructed by retaining dextran-NAD, glutamate dehydrogenase and lactate dehydrogenase in the vicinity of an NH_4^+-sensitive glass electrode positioned behind a dialysis membrane stretched on the tip. Stoichiometric L-alanine production was demonstrated by reductive amination of pyruvate catalysed by alanine dehydrogenase, using galactose-

or lactate dehydrogenase as dextran-NADH regenerating enzymes. A total coenzyme turnover of 90 in 6.5 h in an ultrafiltration membrane reactor (cut-off 10,000) was achieved. In a similar way, accumulation of 6-phosphogluconate and glutamate with nonstoichiometric amounts of dextran-NADP indicated NADP(H) cycling for the first time using a mixture of glucose-6-phosphate- and glutamate dehydrogenase. Okuda et al. [60] later utilized PEG (M_r 3,000)-NADP(H) to confirm the functioning of the same enzyme system in an enzyme membrane reactor.

Other authors have tested various macromolecular NAD derivatives in similar reactor systems, e.g. PEI (M_r 40,000–50,000)-NAD [70] or polylysine (M_r 3,000–85,000)-NAD [84], demonstrating lactate production from pyruvate by lactate dehydrogenase and regenerating polymer-bound NADH with alcohol dehydrogenase using ethanol as substrate. By adding aldehyde dehydrogenase, Yamazaki et al. [84] effected a 20-fold increase in conversion rate for the continuous production of L-lactate. Enzymatic trapping of the acetaldehyde by conversion to acetic acid shifted the equilibrium in favor of L-lactate production and allowed higher space-time yields to be achieved. In both cases, the unacceptable instability of the C(6)-N-acyl-bond of polymer-coupled N^6-succinyl-NAD was reported. Marconi et al. [124] used an elegant method for the entrapment of PEI-NAD (synthesized according to Zappelli et al. [61]), lactate dehydrogenase and alanine dehydrogenase in cellulose triacetate fibers through which only low-molecular-weight compounds could pass. L-alanine was produced for 15 d from pyruvate by the fiber reactor at a rate of 220 µMol h^{-1} per g enzyme fiber, with recycling of PEI-NAD(H) at 222 times per h. At the end of this period, 100 mMol L-alanine was produced per g fiber. The macromolecular NAD(H) was recycled with a TTN of 80,000, the highest cycle number at that time (1975).

A novel membrane reactor configuration was developed by Chang et al. [118], for the biomedical application of enzymes (artificial cells for extracorporal perfusion, enzyme therapy). Spherical microcapsules (mean diameter 5–100 µm) with semiper-meable collodion- or nylon-polyethylene imine membranes, each containing enzymes and coenzymes and representing a mini-enzyme membrane reactor, were utilized in batch mode or column form. These microencapsulated (co)enzyme systems have the advantage of being in an immobilized form, but still in a solubilized state, and, thus free to move and interact inside the microcapsule without steric or diffusion restrictions. The model system malate dehydrogenase/alcohol dehydrogenase/dextran (M_r 70,000)-NAD(H) for the continuous conversion of oxaloacetate to L-malate and the NADH regenerating enzymatic oxidation of ethanol to acetaldehyde was studied. Using col-lodion microcapsules containing tediously purified enzyme-free hemoglobin [125] or easier preparable nylon microcapsules with enclosed polyethylene imine [126], it was demonstrated that both additions could significantly stabilize the enzyme system. Continuous L-malate production (60% conversion of 5 mM oxaloacetate) and an acceptable dextran-NADH regeneration efficiency (90–100 cycles per h) were main-tained over prolonged periods (7 d) to give a TTN of around 16,000.

Recently, Ilan and Chang [32] reported the preparation of semipermeable micro-capsules with ultrathin lipid-polyamide membranes containing L-glutamate dehy-drogenase and alcohol dehydrogenase for the continuous enzymatic trapping of NH_4^+ ions by conversion of α-ketoglutarate in L-glutamate, utilizing nondiffusing native NADH as regenerable coenzyme. This work may be useful in the development of ultrafiltration reactors with retainable native coenzymes.

Fuller et al. [89] were the first to show in batch that polymer-bound NAD(H) could be repeatedly used in a coupled substrate enzyme system with horse liver alcohol dehydrogenase, which catalysed both the NAD-dependent conversion of benzyl alcohol to benzaldehyde and the regeneration of polymer-bound NAD by converting acetaldehyde to ethanol. By monitoring the accumulation of chromogenic benzaldehyde, 600 to 1,600 coenzyme cycles per h were determined, depending on the coenzyme concentration.

With the aim of reducing NAD costs in continuous-flow systems for biochemical analysis, Sakaguchi et al. [127] used dextran-bound NAD in excess for the fluorometric determination of L-lactate or L-glutamate utilizing column immobilized lactate dehydrogenase or glutamate dehydrogenase. Partially reduced dextran-NAD ($<2\%$) was continuously regenerated using the redox compound phenazin methosulfate. This was inserted into the flow system and a fast dialysis step was then used to remove excess redox reagent before reentering the column.

These examples illustrate that macromolecular coenzyme derivatives may be utilized with some versatility in various systems. The commercially available enzymes involved were more of analytical interest and, except for alcohol dehydrogenase, of less importance for synthetic purposes.

A considerable part of the cooperative project on enzymatic synthesis between the GBF (Dept. of Enzyme Technology), KFA Jülich (Institute of Biotechnology) and DEGUSSA-AG in the period 1978–1986 comprised the evaluation of the merits of NAD(H)-dependent enzyme systems and ultrafiltration membrane reactor technology for the synthesis of fine chemicals difficult to prepare by pure organo-chemical methods. The project was set up with particular emphasis on a) the search for valuable enzymes from microbial sources with optimization and scale-up of their production, b) the development of methods for the synthesis of well-characterized coenzymatically active macromolecular NAD(H) derivatives, easily adaptable to larger-scale processes, c) small-scale enzyme membrane reactor experiments for process optimization using chemical engineering principles and d) development of ultrafiltration technology for fine chemical production with promising enzyme/polymer-bound NAD(H) systems on pilot or production scale. New NAD(H)-dependent enzymes were found which catalysed reactions of potential interest: secondary alcohol dehydrogenase [39] (highly active enzyme for oxidation or reduction of secondary alcohols or ketones with broad substrate specificity), L- and D-α-hydroxyisocaproate dehydrogenase [128,129] (enantiospecific reduction of α-ketocarboxylic acids to the corresponding α-hydroxycarboxylic acids with broad substrate specificity), D-mandelate dehydrogenase [130] (reduction of α-ketophenyl acetic acid to D-mandelate) and L-phenylalanine dehydrogenase [16] (highly active enzyme, catalysing the transformation of α-ketophenylpyruvate and α-keto-4-hydroxyphenylpyruvate to, respectively, L-phenylalanine and L-tyrosine by reductive amination with NH_4^+). For these enzymes and other enzymes known from the literature, e.g. L-alanine dehydrogenase, L-leucine dehydrogenase and formate dehydrogenase, work was undertaken to optimize and scale up the bioprocesses and the down-stream processing techniques. The latter was based on a partition in aqueous-two phase systems [136] used to obtain technically feasible enzyme preparations. These investigations were particularly successful for L-leucine dehydrogenase [131], L-phenylalanine dehydrogenase [16,132], formate dehydrogenase [47,48] and L-hydroxyisocaproate dehydrogenase [133].

Table 5. The production of L-leucine, L-*tert*-leucine and L-phenylalanine in an enzyme membrane reactor (10 ml reactor volume), applying L-leucine dehydrogenase (Leu DH, *Bacillus cereus*) and L-phenylalanine dehydrogenase (Phe DH, *Rhodococcus* spec. M4) and using formate dehydrogenase (FDH, *Candida boidinii*) for the regeneration of the macromolecular coenzyme derivative PEG (M_r 20,000)-N^6-(2-aminoethyl)-NADH

Amino acid (product)	Substrates	Conc. [M]	PEG NADH [mM]	Enzymes	Conc. [U ml⁻¹]	Enzyme consumption [U kg⁻¹ Product]	Space-time yield [g l⁻¹ d⁻¹]	Cycle number
L-leucine	α-Keto-iso-caproate	0.1	1.0	Leu DH	5	300	250	100 000
				FDH	5	300		
	Ammonium formate	0.3						
L-tertiary leucine	3,3-Dimethyl-2-keto-butyric acid	0.5	0.4	Leu DH	30	1000	640	130 000
				FDH	30	2000		
	Ammonium formate	1.0						
L-phenyl-alanine	Phenyl-pyruvate	0.12	0.4	Phe DH	20	1500	456	600 000
				FDH	2.5	150		
	Ammonium formate	1.0						

Polyethylene glycol (PEG) was introduced, for the first time, as a water-soluble polymer for increasing the molecular weight of coenzymes [66]. This idea was later taken over by Furukawa et al. [111] who used aminated PEG (M_r 3,000) for the coupling of N^6-(2-carboxyethyl)-NAD.

A simple method for introducing terminal carboxyl groups in PEG (M_r 20,000) was developed and adapted to large-scale synthesis [90]. Two new simplified methods, both adaptable to large-scale usage, for the preparation of N^6-(2-aminoethyl)-NAD(P) [76,137] and technical grade PEG (M_r 20,000)-N^6-(2-aminoethyl)-NADH [90] were developed, as described in Sects. 3.2.1 and 3.2.2.

The optimal operating conditions for the production of L-amino acids by various enzyme/PEG-(M_r 20,000)-N^6-(2-aminoethyl)-NADH combinations were derived from the data of small-scale membrane reactor experiments (10-ml reactor volume), leading to impressive space/time yields of the order of 250–640 $gl^{-1} d^{-1}$, e.g. for L-leucine, L-tert-leucine and L-phenylalanine [134].

Significantly lower space/time yields are usually obtained for bioprocesses utilizing whole cells. Some data are summarized in Table 5 [134] for the production of these three L-amino acids. As an example, the long-term enzymatic conversion of 3,3-dimethyl-2-ketobutyric acid to L-tert-leucine is shown in Fig. 15, using a flat-bed enzyme membrane reactor with a reactor-volume of 10 ml according to the reaction

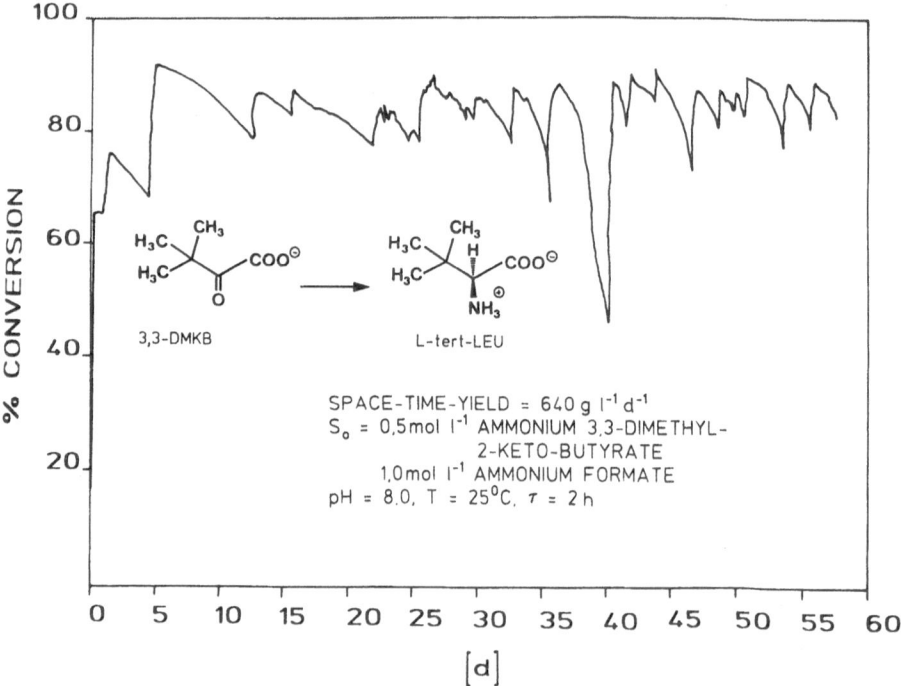

Fig. 15. Long-term enzymatic conversion of 3,3-dimethyl-2-ketobutyric acid (3,3-DMKB) to L-tert-leucine (L-tert-leu) with the leucine dehydrogenase (Leu DH)/formate dehydrogenase (FDH)/PEG (M_r 20,000)-N^6-(2-aminoethyl)-NAD(H) system (for enzyme- and coenzyme concentrations see Table 5)

τ: residence time

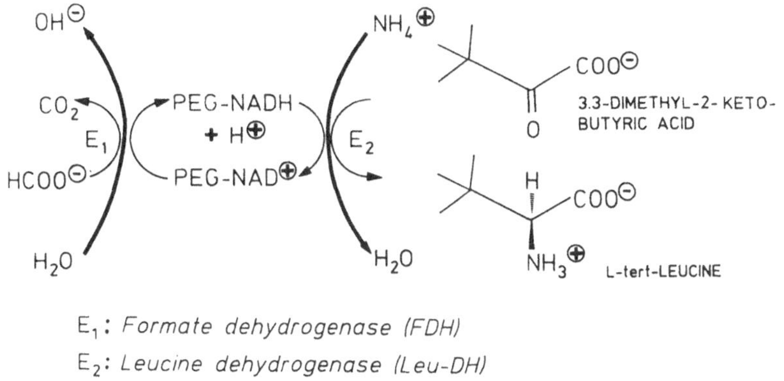

E_1: Formate dehydrogenase (FDH)

E_2: Leucine dehydrogenase (Leu-DH)

Fig. 16. Reaction pathway for the enzymatic synthesis of L-*tert*-leucine from 3,3-dimethyl-2-keto-butyric acid with simultaneous enzymatic NADH regeneration

pathway of Fig. 16. The more-or-less sharp-edged character of the conversion course stems from the periodic addition of enzymes and/or PEG-(M_r 20,000)-N^6-(2-aminoethyl)-NADH to maintain a conversion of around 80%. Since very high NAD(H) cycle numbers in the range 100,000–600,000 (TTN, expressed as mol product formed per mol NAD(H) consumed during the total reaction time) could be obtained (Table 5), coenzyme usage was not the limiting factor from the economic point of view.

Nevertheless, the prohibitive high cost of the α-keto acids as substrates, e.g. α-keto-isocaproic acid or phenylpyruvate, is the main bottleneck of the enzyme membrane reactor concept, where NAD(H)-dependent enzymes and macromolecular NAD(H) derivatives are used for the large-scale production of the natural amino acids L-leucine or L-phenylalanine. To date, bioprocesses utilizing whole cells converting cheap substrates are more economical.

Apart from the successful development of new methodology for the preparation of partially purified technically relevant enzymes, N^6-(2-aminoethyl)-NAD(P) and PEG-(M_r 20,000)-N^6-(2-aminoethyl)-NADH, the project also showed that large-scale NAD(H)-dependent enzymatic synthesis of useful chiral compounds was possible. L-leucine dehydrogenase from *Bacillus cereus* with substrate specificity for natural hydrophobic branched-chain L-amino acids and the corresponding α-ketocarboxylic acids turned out to be active with respect to the stereospecific reductive amination of 3,3-dimethyl-2-ketobutyric acid to the unnatural amino acid L-*tert*-leucine [V_{max} (3,3-dimethyl-2-ketobutyric acid) = 5.2 U mg^{-1} versus V_{max} (α-ketoisocaproate) = 11.5 U mg^{-1}, K_M (3,3-dimethyl-2-ketobutyric acid) = 37.5 mM, unfavorable compared to K_M (α-ketoisocaproate) = 0.02 mM, using PEG (M_r 20,000)-N^6-(2-aminoethyl)-NADH in excess] [14].

Chemical methods exist that yield the DL-racemate of L-*tert*-leucine. These, however, require a rather tedious chemical or enzymatic step for isolation of the pure L-enantiomer.

Since L-*tert*-leucine is an unnatural compound, bioprocesses analogous to the production of natural L-amino acids from simple substrates are not possible. However, for the direct synthesis of L-*tert*-leucine in kg quantities, similar operational conditions were maintained to those outlined in Fig. 15 and Table 5 with respect to residence time and concentration of leucine dehydrogenase, formate dehydrogenase, PEG-(M_r 20,000)-N^6-(2-aminoethyl)-NADH (now of technical grade) and substrates for applying a flatbed membrane reactor (cut-off 5,000) with a reactor volume of approx. 1.5 l. Up to 1 kg L-*tert*-leucine could be produced per day [139]. It should be emphasized that due to the less favorable kinetics of leucine dehydrogenase with respect to 3,3-dimethyl-2-ketobutyric acid as unnatural substrate, sufficient continuous transformation to L-*tert*-leucine could be maintained by raising the concentration of this very soluble acid to 0.5 M [approx. $14 \times K_M$ (3,3-dimethyl-2-ketobutyric acid)], the NH_4^+ concentration to 1 M (2.5-fold) and both enzyme levels (6-fold), while leaving the PEG-NAD(H) concentration unchanged compared to optimal production conditions for L-leucine in model experiments [14].

These reaction conditions can be used for scale-up: e.g. a 200-l membrane reactor would be able to produce approx. 130 kg L-*tert*-leucine per day under similar conditions (residence time 2 h: flow rate 100 l h^{-1}) with 6×10^6 U leucine dehydrogenase and formate dehydrogenase and 80 mMol PEG (M_r 20,000)-NAD(H) (approx. 1.6 kg) permanently present in the reactor. Since leucine dehydrogenase and formate dehydrogenase can be produced at levels of 4,000 U and 1,000 U l^{-1}, respectively, an acceptable fermentor capacity is required of 2×10^3 and 8×10^3 l, taking into account an enzyme yield of 70 % after partial purification by the aqueous two phase method that results in technical grade enzyme preparations [47,48]. The amount of PEG (M_r 20,000)-N^6-(2-aminoethyl)-NADH required might be conveniently synthesized as described by Bückmann et al. [90].

It should be emphasized that L-*tert*-leucine is the first commercially attractive compound synthesized by coenzyme-dependent enzyme systems with simultaneous enzymatic coenzyme regeneration, using a bioreactor based on ultrafiltration (enzyme membrane reactor) applying a macromolecular coenzyme derivative.

The versatility of unnatural L-*tert*-leucine as a chiral auxiliary for the synthesis of rare amino acids and diverse chiral aldehydes, ketones and olefins is illustrated in Fig. 17. Usually higher enantiomeric excess values are induced compared to the commonly used L-valine [138] by applying L-*tert*-leucine with a bulkier *tert*-butyl side group.

Furthermore, L-*tert*-leucine is a building block of the peptide antibiotic bottromycin and has been employed as substitute for L-leucine or other L-amino acids in modified peptides (enkaphelin, vasopressin, oxytocin and dalargin analogues) and as a chiral phase in HPLC and GC [138]. L-*tert*-leucinol, prepared by chemical reduction of the carboxyl group of L-*tert*-leucine, is a valuable chiral auxiliary for the synthesis of rare chiral ketones, carboxylic acids and α-hydroxy acids [138].

If α-$^{15}NH_2$-labeled L-amino acids become important for future applications, the (co)enzyme system L-amino acid dehydrogenase/formate dehydrogenase/PEG (M_r 20,000)-N^6-(2-aminoethyl)-NADH offers a convenient method for their synthesis, since the preparation of a broad spectrum of L-amino acids (up to kg amounts) might be covered by L-leucine-, L-alanine-, L-glutamate-, and L-phenylalanine dehydrogenase.

As yet, no continuous synthesis of valuable fine chemicals by applying NADP(H)-dependent enzyme systems using macromolecular NADP(H) derivatives have been

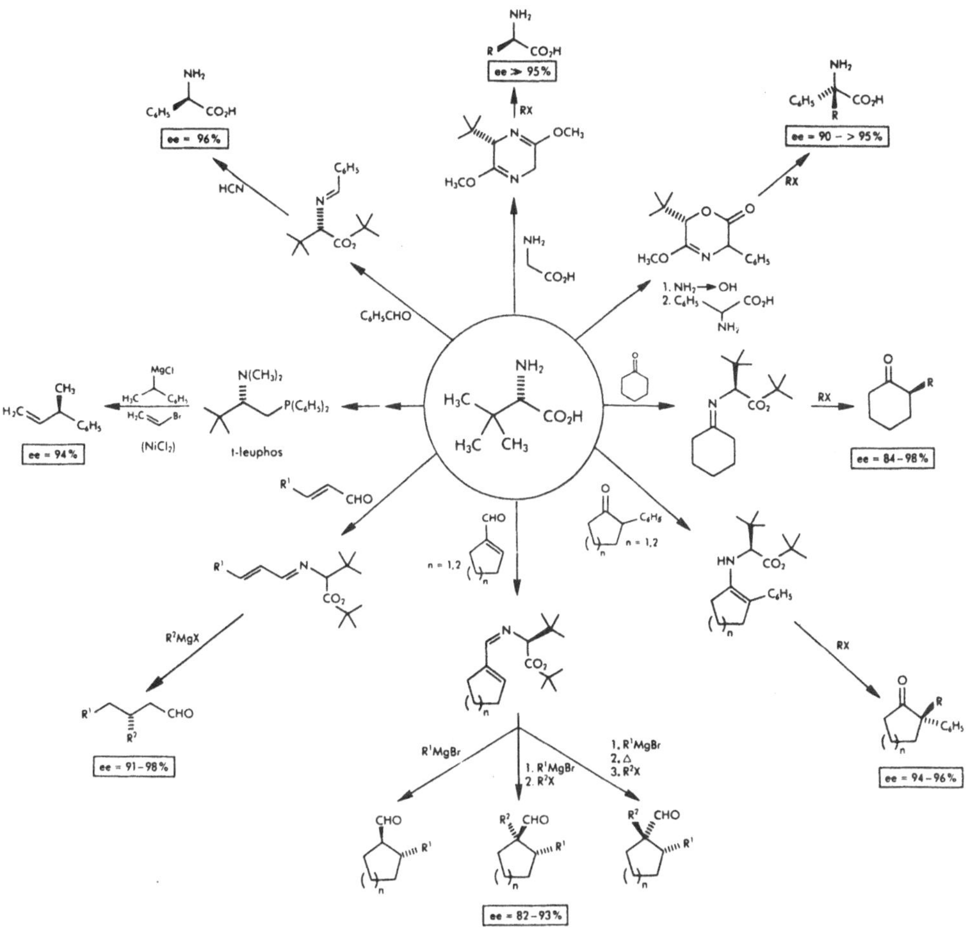

Fig. 17. Versatility of L-*tert*-leucine as chiral (auxiliary [138];) ee: enantiomeric excess

reported. The enzyme system 12α-hydroxysteroid dehydrogenase/glutamate dehydro-genase for the NADP-dependent conversion of cheap cholic acid to expensive 12-keto-chenodeoxycholic acid with simultaneous NADP regeneration by glutamate dehydrogenase [20] might have been an attractive application if macromolecular NADP derivatives had shown satisfactory coenzyme activity with 12α-hydroxysteroid dehydrogenase (Sect. 4.3).

The potential of active macromolecular FAD derivatives functioning as FAD reservoirs for oxidases has already been considered in Sect. 4.3.

The evaluation of the merits of artificial enzyme-FAD conjugates with FAD covalently coupled via the adenine, synthesized after preparing apoenzymes from holoenzymes with noncovalently bound FAD, might be an interesting subject for further study.

6 Future Outlook

The utilization of macromolecular coenzyme derivatives in connection with ultra-filtration reactor technology, e.g. the enzyme-membrane reactor, for continuous transformations by coenzyme-dependent enzymes with simultaneous coenzyme regeneration can be cosidered an acceptable reactor concept. It should be emphasized that this concept is not yet of routine character, since the application of immobilized cells is a competitive alternative that will facilitate the coenzyme regeneration. Favorable conditions for a more extensive application of systems with enzymes and macromolecular coenzyme derivatives in the future will be the development of production methods of further coenzyme-dependent enzyme activities of catalytic potential from high-expression microorganisms obtained by classic mutation or recombinant DNA techniques. The presence of simple methods for the synthesis of suitable macromolecular coenzyme derivatives adaptable to large-scale preparation, as has been exemplified for PEG (M_r 20.000)-N^6-(2-aminoethyl)-NADH of technical grade, will be a further prerequisite.

The existence of NAD(P) (H)-dependent alcohol dehydrogenases with broad substrate specificity has not yet led to their synthetic application in ultrafiltration reactors, although macromolecular NAD(P) (H) derivatives do show satisfactory coenzyme activity for these enzymes. The coupled substrate approach for coenzyme regeneration without the need of a second enzyme may be interesting to study with regard to the enzyme membrane reactor concept using macromolecular coenzyme derivatives and alcohol dehydrogenase. The evaluation of the merits of enzyme-coenzyme conjugates, obtained by covalently binding a reactive coenzyme derivative by more rational methods will be an important subject of study. The properties of these conjugates, functioning as coenzyme-dependent enzymes and as macromolecular coenzyme derivatives, must be thoroughly investigated to see, for instance, to what extent the anchimeric assistance effect, a consequence of the intramolecular reaction inside enzyme-coenzyme conjugates, can be exploited with respect to the application of coenzyme-dependent enzymes.

7 Acknowledgements

The experimental work of the authors was partly supported by a grant from the biotechnology program of the Bundesministerium für Forschung und Technologie (BCT 310A) and the biotechnology action program of the Commission of the European Communities (BAP-0060). We thank Prof. C. Wandrey, Institute of Bio-technology, KFA Jülich (FRG), Dr. W. Leuchtenberger and Dr. K.-H. Drauz, DEGUSSA-AG, Hanau/Wolfgang (FRG), for providing the data of Figs. 15 and 16, and Table 5.

8 References

1. Klibanov AM (1983) Technology Review 86: 50
2. Godtfredsen SE, Ingvorsen K, Yde B, Andresen O (1985) The scope of biocatalysts in organic chemical processing, in: Studies in organic chemistry 22, Biocatalysts in organic syntheses; Tramper J, van der Plas HC, Linko P (eds) Amsterdam, Elsevier, p 3

3. Sugiura Y, Kuwahara J, Nagasawa T, Yamada H (1987) J. Am. Chem. Soc. 109: 5848
4. Linko YY, Linko P (1985) Immobilized biocatalysts in organic synthesis and chemical production, in: see Ref. 2, p 159
5. Laane C, Tramper J, Lilly MD (eds) (1987) Studies in organic chemistry 29, Biocatalysis in organic media, Amsterdam, Elsevier
6. Chenault HK, Whitesides GM (1987) Appl. Biochem. Biotech. 14: 147
7. Whitesides GM, Wong CH (1985) Angew. Chem. 97: 617
8. Matos JR, Smith MB, Wong CH (1985) Bioorg. Chem 13: 121
9. Jones JB (1986) Tetrahedron 42: 3351
10. Keinan E, Hafeli EK, Seth KK, Lamed R (1986) J. Am. Chem. Soc. 108: 162
11. Lee LG, Whitesides GM (1986) J. Org. Chem. 51: 25
12. Simon H, Bader J, Günther H, Neumann S, Thanos J (1985) Angew. Chem. 97: 541
13. Wichmann R, Wandrey C, Bückmann AF, Kula MR (1981) Biotech. Bioeng. 23: 2789
14. Wandrey C, Bossow B (1985) Continuous Cofactor Regeneration, in: Proc. 3rd Int. Conf. Chemistry and Biotechnology of Biologically Active Natural Products, Sept. 16–21, 1985, Sofia (Bulgaria) Bulgarian Acad. Sci. 1: 195
15. Ohshima T, Wandrey C, Kula MR, Soda K (1985) Biotech. Bioeng. 27: 1616
16. Hummel W, Schütte H, Schmidt E, Wandrey C, Kula MR (1987) Appl. Microbiol. Biotechnol. 26: 409
17. Asano Y, Nakazawa A (1987) Agric. Biol. Chem. 51: 2035
18. Asano Y, Endo K, Nakazawa A, Hibino Y, Okazaki N, Ohmori M, Numao N, Kondo K (1987) Agric. Biol. Chem. 51: 2621
19. Hummel W, Schütte H, Kula MR (1984) Enzyme Engineering 7; Ann. N. Y. Acad. Sci. 434: 87
20. Riva S, Bovara R, Pasta P, Carrea G (1986) J. Org. Chem. 51: 2902
21. Crans DC, Kazlauskas RJ, Hirschbein BL, Wong CH, Abril O, Whitesides GM (1987) Methods in Enzymology, 136: 263
22. May SW, Padgette SR (1983) Bio/Technology, 5: 677
23. Alberti BN, Klibanov AM (1982) Enzyme Microb. Technol. 4: 47
24. Szwajcer E, Brodelius P, Mosbach K (1982) Enzyme. Microb. Technol. 4: 409
25. Aretz W, Sauber K (1984) L-Amino Acid Oxidase, in: Proc. 3rd Europ. Conf. Biotechnology, September 10-14, 1984, München (FRG), Vol. I, Rehm HJ, Behrens D (eds) Weinheim, Verlag Chemie, p 445
26. Tu SC, Edelstein SJ, McCormick DB (1973) Arch. Biochem. Biophys. 159: 880
27. Murata K, Tani K, Kato J, Chibata I (1979) J. Appl. Biochem. 1: 282
28. Dennda G, Kula MR (1986) Biotechnol. 4: 143
29. Wang SS, King CK (1979) Adv. Biochem. Eng. 12: 119
30. Lee LG, Whitesides GM (1985) J. Am. Chem. Soc. 107: 6999
31. Knowles C (1983) Patent Application PCT/GB83/00175
32. Ilan E, Chang TMS (1986) Appl. Biochem. Biotech. 13: 221
33. Kitpreechavanich V, Nishio N, Hayashi M, Nagai S (1985) Biotech. Lett. 7: 657
34. Kulbe KD, Chmiel H (1988) Enzyme Engineering 9; Ann. N.Y. Acad. Sci. 542: 444
35. Mosbach K (1978) Adv Enzymology, 46: 205
36. Katayama N, Hayakawa K, Urabe I, Okada H (1984) Enzyme Microb. Technol. 6: 538
37. Everse J, Zoll EL, Kahan L, Kaplan NO (1971) Bioorg. Chem. 1: 207
38. Chambers RP, Walle EM, Baricos WH, Cohen W (1978) Enzyme Engineering, 3: 363
39. Schütte H, Hummel W, Kula MR (1982) Biochem. Biophys. Acta 716: 298
40. Wong CH, Whitesides GM (1981) J. Am. Chem. Soc. 103: 4890
41. Wong CH, Drueckhammer DG, Sweers HM (1985) J. Am. Chem. Soc. 107: 4028
42. Bückmann AF (1987) Progress Report 1987, Biotechnology Action Programme of the EEC, 2: 255
43. Kazlauskas RJ, Whitesides GM (1985) J. Org. Chem. 50: 1069
44. Wong CH, McCurry SD, Whitesides GM (1980) J. Am. Chem. Soc. 102: 7983
45. Wandrey C, Wichmann R, Leuchtenberger W, Kula MR, Bückmann AF (1982) US Patent 4.304.858
46. Wandrey C, Wichmann R, Leuchtenberger W, Kula MR, Bückmann AF (1982) US Patent 4.326.031

47. Kroner KH, Schütte H, Stach W, Kula MR (1982) J. Chem. Tech. Biotechnol. 32: 130
48. Cordes A, Kula MR (1986) J. Chromatogr. 376: 375
49. Yamamoto I, Saiki T, Liu SM, Ljungdahl LG (1983) J. Biol. Chem. 258: 1826
50. Rella R, Raia CA, Pisani FM, Trincone A, Vaccaro C, Nucci R, Gambacorta A, De Rosa M, Rossi M (1987) Specifity and Stability in Organic Solvents of a Novel Archaebacterial NAD-Dependent Alcohol Dehydrogenase, in: Proc. 4th Europ Congr on Biotechn, June 14–19, 1987, Amsterdam (NL), Vol. 2, Neijssel OM, van der Meer RR, Luyben KChAM (eds) Amsterdam, Elsevier, p 336
51. Rossmann MG, Moras D, Olsen KW (1974) Nature 250: 194
52. Wierenga RK, Drenth J, Schulz GE (1983) J. Mol. Biol. 167: 725
53. Mansson MO, Mosbach K (1987) in: Coenzymes and Cofactors, Dolphin D, Paulson R, Avramovic O (eds) 2B: 217, Wiley, New York
54. Okada H, Urabe I (1987) Methods in Enzymology 136: 34
55. Sogin DC (1976) J. Neurochem. 27: 1333
56. Jones JB, Taylor KE (1976) Can. J. Chem. 54: 2969
57. Lindberg M, Larsson PO, Mosbach K (1973) Eur. J. Biochem. 40: 187
58. Lowe CR, Mosbach K (1974) Eur. J. Biochem. 49: 511
59. Muramatsu M, Urabe I, Yamada Y, Okuda H (1977) Eur. J. Biochem. 80: 111
60. Okuda K, Urabe I, Okada H (1985) Eur. J. Biochem. 151: 33
61. Zappelli P, Rossodivita A, Re L (1975) Eur. J. Biochem. 54: 475
62. Zappelli P, Pappa R, Rossodivita A, Re L (1977) Eur. J. Biochem. 72: 309
63. Zappelli P, Pappa R, Rossodivita A, Re L (1978) Eur. J. Biochem. 89: 491
64. Schmidt HL, Grenner G (1976) Eur. J. Biochem. 67: 295
65. Weibel MK, Fuller CW, Stadel JM, Bückmann, AF, Doyle T, Bright HJ (1974) Enzyme Engineering 2: 203
66. Bückmann AF (1979) German Patent DP 28.41.414
67. Sakamoto H, Nukamura A, Urabe I, Yamada Y, Okada H (1986) J. Ferment. Technol. 64: 511
68. Bückmann AF (1988) Heterocycles 27: 1623
69. Imahori K, Tomita K (1980) German Patent DP 2945129.4
70. Wykes JR, Dunnill P, Lilly MD (1972) Biochim. Biophys. Acta 286: 260
71. Lee CY, Kaplan NO (1975) Arch. Biochem. Biophys. 168: 665
72. Zappelli P, Rossodivita A, Prosperi G, Pappa R, Re L (1976) Eur. J. Biochem. 62: 211
73. Zappelli P, Pappa R, Rossodivita A, Re L (1977) Eur. J. Biochem. 72: 309
74. Okuda K, Urabe I, Okada H (1985) Eur. J. Biochem. 147: 249
75. Riva S, Carrea G, Veronese FM, Bückmann AF (1986) Enzyme Microb. Technol. 9: 556
76. Bückmann AF (1987) Europ Patent Appl 0 247 537.A2
77. Bückmann AF, Wray V, van der Plas HC (1987) Two Unexpected Transformations of N(1)-(2-Aminoethyl)-Adenine Derivatives of NAD, NADP and FAD under Mild Aqueous Conditions, in: Proc. 11th Intl. Conf. Heterocyclic Chemistry, August 16–21, 1987, Heidelberg (FRG) Neidlein R (ed) Frankfurt, GDCh, p 398
78. Davies P, Mosbach K (1974) Biochim. Biophys. Acta 370: 329
79. Malinauskas AA, Kulys JJ (1978) Biotech. Bioeng. 20: 769
80. Sakaguchi Y, Murachi T (1980) J. Appl. Biochem. 2: 117
81. Bückmann AF, Kula MR, Wichmann R, Wandrey C (1981) J. Appl. Biochem. 3: 301
82. Katayama N, Urabe I, Okada H (1983) Eur. J. Biochem. 132: 403
83. Hayakawa K, Urabe I, Okada H (1985) J. Ferment. Technol. 63: 245
84. Yamazaki Y, Maeda H, Suzuki H (1976) Biotech. Bioeng. 18: 1761
85. Yoshikawa M, Goto M, Ikura K, Sasaki R, Chiba H (1983) Agric. Biol. Chem. 46: 207
86. Bückmann AF, Morr M, Johansson G (1981) Makromol. Chem. 82: 1379
87. Yamazaki Y, Maeda H (1981) Agric. Biol. Chem. 45: 2277
88. Le Goffic F, Sicsic S, Vincent C (1980) Eur. J. Biochem. 108: 143
89. Fuller CW, Rubin JR, Bright HJ (1980) Eur. J. Biochem. 103: 421
90. Bückmann AF, Morr M, Kula MR (1987) Biotechnol. Appl. Biochem. 9: 258
91. Albertsson PA (1985) Partition of cell particles and macromolecules, 3rd ed., Wiley Interscience, New York
92. Mansson MO, Larsson PO, Mosbach K (1987) Eur. J. Biochem. 86: 455

93. Mansson MO, Larsson PO, Mosbach K (1979) FEBS Lett. 98: 309
94. Gacesa P, Venn RF (1979) Biochem. J. 177: 369
95. Kovar J, Simek K, Kucera I, Matyska L (1984) Eur. J. Biochem. 139: 585
96. Schäfer HG, Jacobi T, Eichhorn H, Woenckhaus C (1986) Biol. Chem. Hoppe-Seyler 367: 969
97. Eguchi, T, Iizuka T, Kagotani T, Lee JH, Urabe I, Okada H (1986) Eur. J. Biochem. 155: 415
98. Nakamura A, Urabe I, Okada H (1986) J. Biol. Chem. 261: 16792
99. Kato N, Yamagami T, Shimao M, Sakazawa C (1987) Appl. Microb. Biotechnol. 25: 415
100. Goulas P (1987) Eur. J. Biochem. 168: 469
101. Woenckhaus C, Koob R, Burkhard A, Schäfer HG (1983) Bioorg. Chem. 12: 45
102. Jacobi T, Woenckhaus C (1987) Binary enzyme reactors with modified dehydrogenase, in: Proc. 4th Europ Congr Biotechn, Vol. 1 Neijssel OM, van der Meer RR, Luyben KChAM (eds) Elsevier, Amsterdam, p 72
103. Wandrey C, Wichmann R (1985) Coenzyme regeneration in membrane reactors, in: Application of isolated enzymes and immobilized cells to biotechnology Laskin A (ed) Addison Wesley, p 177
104. Adachi S, Ogata M, Tobita H, Hashimoto K (1984) Enzyme Microbiol. Technol. 6: 259
105. Grenner G, Schmidt HL, Völkl W (1976) Hoppe-Seyler's Z. Physiol. Chem. 357: 887
106. Bränden CJ, Jornvall H, Eklund H, Furugren B (1975) Alcohol dehydrogenase, in: The Enzymes, Vol. 11, Boyer PD (ed) New York, Academic Press p 103
107. Holbrook JJ, Liljas A, Steindl SJ, Rossmann, MG (1975) Lactate dehydrogenase, in: The Enzymes, Vol. 11, Boyer PD (ed), New York, Academic Press p 191
108. Yamazaki Y, Maeda H, Satoh A, Hiromi K (1984) J. Biochem. 95: 109
109. Schmidt HL, Dolabdjian B (1980) Methods in Enzymology 66: 176
110. Furukawa D, Urabe I, Okada H (1981) Eur. J. Biochem. 114: 101
111. Furukawa S, Katayama N, Iizuka T, Urabe I, Okada H (1980) FEBS Lett. 121: 239
112. Hummel W, Schütte H, Kula MR (1984) Enzyme Engineering 7; Ann. N. Y. Acad. Sci. 434: 194
113. Jones JB (1985) An illustrative example of a synthetically useful enzyme: Horse liver alcohol dehydrogenase, in: Enzymes in organic synthesis, Ciba Foundation Symposium 111, Porter R, Clark S (eds) London, Pitman, p 3
114. Whitesides GM (1985) Applications of cell-Free enzymes in organic synthesis, in: Enzymes in organic synthesis, Ciba Foundation Symposium 111, Porter R, Clark S (eds) London, Pitman, p 76
115. Carrea G, Bovara R, Longhi R, Riva S (1985) Enzyme Microb. Technol. 7: 597
116. Kajiwara S, Maeda H (1985) Biotech. Bioeng. 18: 1794
117. Yamazaki Y, Maeda H (1982) Agric. Biol. Chem. 46: 1571
118. Chang TMS (1987) Methods in Enzymology 16: 67
119. Flaschel E, Wandrey C, Kula MR (1983) Adv. Biochem. Eng/Biotechnol. 26: 73
120. Fiolitakis E, Wandrey C (1982) Reaction technology of the enzymatically catalysed production of L-alanine, in: Proc. 3rd Rotenburg Fermentation Symposium, Lafferty RM (ed) p 273
121. Leuchtenberger W, Wandrey C, Kula MR (1986) German Patent, 33.07.094.6
122. Tichy S, Vasic-Racki D, Schütte H, Talsky G, Wandrey C (1987) Chem. Biochem. Eng. 1: 25
123. Schmidt E, Vasic-Racki D, Wandrey C (1987) Appl. Microbiol. Biotechnol. 26: 42
124. Marconi W, Prosperi G, Giovenco S, Morisi F (1975) J. Mol. Catal. 1: 111
125. Grunwald J, Chang TMS (1979) J. Appl. Biochem. 1: 104
126. Grunwald J, Chang TMS (1981) J. Mol. Catal. 11: 83
127. Sakaguchi Y, Sukahara M, Endo J, Murachi T (1981) J. Appl. Biochem. 3: 32
128. Schütte H, Hummel W, Kula M-R (1984) Appl. Microbiol. Biotechnol. 19: 167
129. Hummel W, Schütte H, Kula MR (1985) Appl. Microbiol. Biotechnol. 21: 7
130. Hummel W, Schütte H, Kula MR, Leuchtenberger W (1985) German Pat Appl P 35.36.662.1
131. Schütte H, Hummel W, Tsai H, Kula MR (1985) Appl. Microbiol. Biotechnol. 22: 306
132. Campagna R, Bückmann AF (1987) Appl. Microbiol. Biotechnol. 26: 417
133. Schütte H, Hummel W, Kula MR, Leuchtenberger W (1985) German Patent P 32.34.022.2
134. Schmidt E, Wichmann R, Kula MR, Wandrey C (1987) Production of L-amino acids from their corresponding α-ketoacids, in: Proc. of the Sectorial Meeting of Contractors, CEC Bio-

technology Action Programme (BAP), Enzyme Engineering: Protein design and application in biocatalysis, Capri (Italy) Rossi M (ed) Brussels, CEC, p 131

135. Furukawa S, Sugimoto Y, Urabe I, Okada H (1980) Biochimie 62: 629
136. Kula MR, Kroner KH, Hustedt H (1982) Adv. Biochem. Eng. 24: 73
137. Bückmann AF (1987) Biocatalysis 1: 173
138. Info 87-9 (1987) Merck-Schuchardt, Germany (FRG)
139. Lotter H, Drauz KH, Kleemann A, Leuchtenberger W, Wandrey C, Kula MR (1987) Herstellung und Anwendung von L-*tert*-Leucin, Poster presented at the 21th GDCh Hauptversammlung, 13–18 Sept, 1987, Berlin

Enzymes Involved in Penicillin, Cephalosporin and Cephamycin Biosynthesis

Juan F. Martín and Paloma Liras
Universidad de León, Departamento de Microbiología, Facultad de Biología, León, Spain

1 Introduction

Most of the biosynthetic reactions involved in the formation of penicillins and cephalosporins have now been proved in cell-free systems, although there are still some questions remaining about the exact role of some enzymes and intermediates (Fig. 1). A few of these enzymes, e.g. the isopenicillin N synthases of *Penicillium chrysogenum, Cephalosporium acremonium* (syn. *Acremonium chrysogenum*), *Streptomyces clavuligerus* and *S. lactamdurans* (syn. *Nocardia lactamdurans*), the acyltransferase of *P. chrysogenum* and the deacetoxycephalosporin C synthase of *C. acremonium* and *S. lactamdurans* have been highly purified. By contrast, the formation of the tripeptide δ(L-α-aminoadipyl)-L-cysteinyl-D-valine (LLD-ACV) in all β-lactam producing microorganisms and the late enzymes involved in cephamycin biosynthesis

Advances in Biochemical Engineering/
Biotechnology, Vol. 39
Managing Editor: A. Fiechter
© Springer-Verlag Berlin Heidelberg 1989

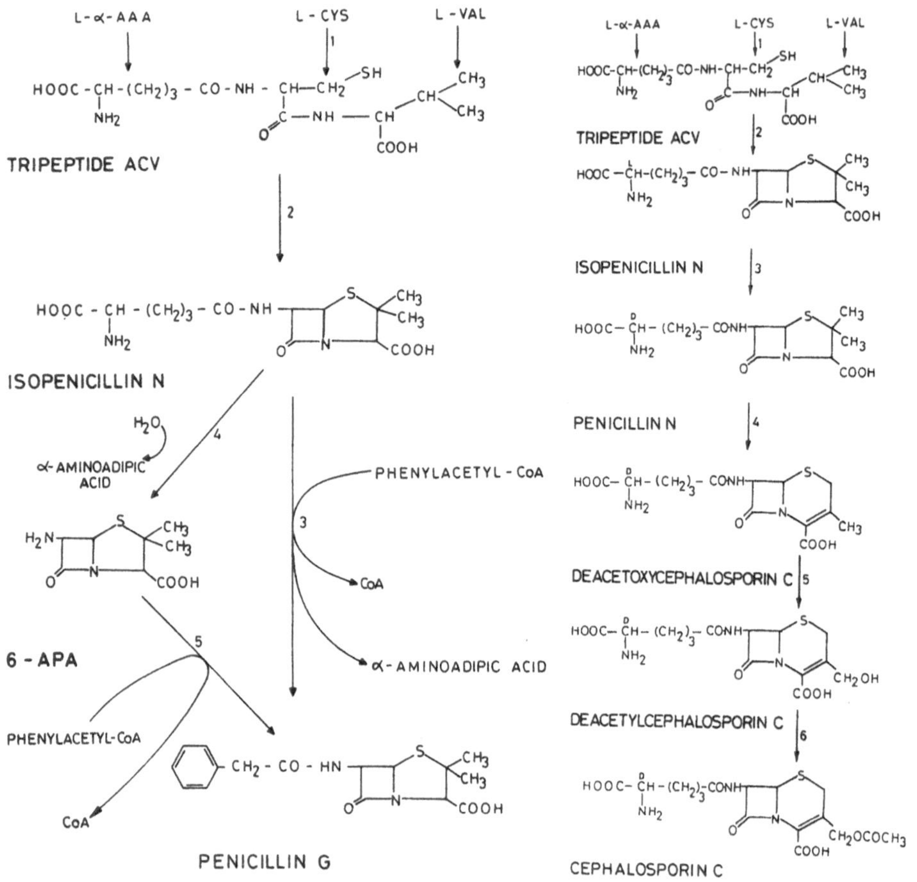

Fig. 1. *Left*: Biosynthetic pathway of penicillin G from the amino acids L-α-aminoadipic, L-cysteine and L-valine. 1, ACV synthetase. 2, Isopenicillin N synthase. 3, Isopenicillin N acyltransferase. 4, Isopenicillin N amidase (6-APA forming). 5, 6-APA acyltransferase. *Right*: Biosynthetic pathway of cephalosporin C from the same component amino acids. 1. ACV synthetase. 2. Isopenicillin N synthase. 3. Isopenicillin N epimerase. 4, Deacetoxycepahlosporin C synthetase. 5, Deacetoxy-cephalosporin C hydroxylase. 6, Deacetylcephalosporin C acetyltransferase. Note that the two initial steps are identical in both biosynthetic pathways

remain largely unknown. Some of these enzymatic reactions involved in the bio-synthesis of penicillins and cephalosporins (e.g. formation of the β-lactam thiazolidine nucleus catalyzed by isopenicillin N synthase and the ring expansion carried out by the deacetoxycephalosporin C synthase) belong to new types of reactions which have little or no parallel in other fields of biochemistry [1].

2 Enzymes Involved in Penicillin and Cephalosporin Biosynthesis

2.1 Biosynthesis of δ(L-α-Aminoadipyl)-L-Cysteinyl-D-Valine

The tripeptide ACV is an intermediate in the biosynthesis of penicillins, cephalosporins and cephamycins. The tripeptide has been isolated and identified unequivocally
from the mycelia of *P. chrysogenum*, *C. acremonium*, *S. clavuligerus*, *S. lactamdurans*
and also in extracts of *Paecilomyces persicinus*, a producer of cephalosporins [2].
ACV is found both in cell extracts and in the medium of high and low penicillin
producing strains [3]. The presence of ACV in filtrates, also observed by Adriaens
et al. [4] is intriguing since such peptides cannot be used in penicillin production because
the cyclization enzyme is intracellular. Early studies related to the characterization of
the tripeptide and to the determination of the stereochemistry have been reviewed [1, 5, 6]
and are not considered here in detail.

There is a good correlation between the tripeptide-forming activity of different
strains and the level of penicillin or cephalosporin synthesized. Formation of ACV
has been studied in our laboratory by following the incorporation of (^{14}C)valine
and (^{14}C)α-aminoadipic acid into ACV [3]. A low penicillin producing strain showed
low ACV-forming activity and also low levels of isopenicillin N synthase (IPNS)
and Acyl-CoA:6-APA acyltransferase as compared to high penicillin producers.
The level of ACV in extracts of high yielding strains of *C. acremonium* obtained from
pharmaceutical companies appeared to be considerable higher than in the strains
available in University laboratories [1]. The availability of such improved strains has
been extremely useful for a better understanding of the enzymes involved in the biosynthesis of β-lactam antibiotics.

ACV formation is clearly stimulated when protein synthesis is inhibited with cyclo-
heximide or anisomicin [3] indicating that this peptide is formed by a nonribosomal
condensation of the three component amino acids as suggested by Fawcett and
Abraham [7]. Blocking protein synthesis may channel amino acids towards nonribosomal ACV formation, as suggested by Katz and Weissbach [8] in the actinomycin
producing strain. In vivo ACV formation is also stimulated by α-aminoadipic acid,
and inhibited by leucine, isoleucine and D-valine [3].

The dipeptide δ(L-α-aminoadypyl)-L-cysteine (AC) has been isolated from cultures
of *C. acremonium* [9], which suggests that the reaction starts with the activation of
the L-α-aminoadipic acid residue at its δ-carboxyl (Fig. 2) followed by activation of
the cysteine and the reaction of the α-amino group of L-cysteine (or an activated form
of L-cysteine) with the δ-carboxyl group of activated α-aminoadipic acid to yield the
dipeptide L-α-aminoadipyl-L-cysteine (or an activated form of the same dipeptide) [10].
The LL-AC dipeptide is converted into the LLD-ACV by reaction of the α-amino
group of L-valine with an activated form of the dipeptide. It seems likely that the
L-isomer of valine is epimerized while it is enzyme bound as part of the second step
of the formation of the tripeptide. Studies on the incorporation of (^{3}H)valine and
(^{14}C)valine into penicillin are all consistent with the occurrence of an enzyme bound
form of L-valine during conversion of L-valine into LLD-ACV. Furthermore, one of
the two carboxy oxygen atoms of L-(^{18}O$_2$)valine was removed during the conversion
of AC and valine into ACV which also supports the model of an enzyme bound

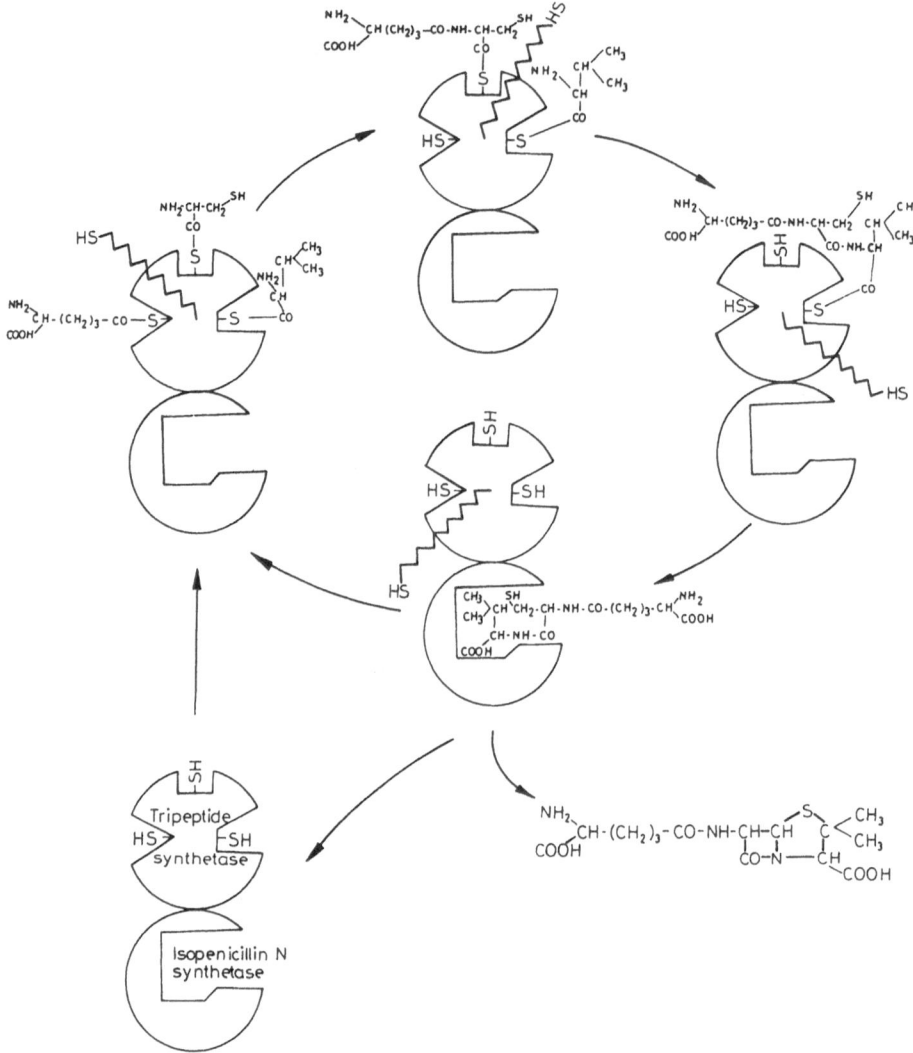

Fig. 2. Proposed model of ACV formation by the ACV synthetase (upper circle) starting with the activation of L-α-aminoadipic acid (left), followed by activation of the L-cysteine and L-valine and the transfer of the activated L-α-aminoadipic acid to form the dipeptide α-aminoadipyl-L-cysteine (top). Finally the dipeptide is transferred onto D-valine to form the tripeptide LLD-ACV. The involvement of a pantetheinyl residue as a swinging arm in the transfer of the activated amino acids has not been established. Most of the tripeptide appears to be cyclized very efficiently by the isopenicillin N synthase (lower circle) and only a minor ACV part is lost without cyclization to the medium (Ref. [3])

intermediate [11]. The activated forms of the amino acids and the dipeptide have still to be determined.

The order in which the amino acid residues of ACV are coupled was examined initially by Abraham and co-workers. Cell-free extracts of *C. acremonium* in the presence of an enzymatic ATP generating system converted a mixture of δ(L-α-aminoadypyl)-L-cysteine and (^{14}C)valine to radioactive ACV, but the same extracts failed to convert a mixture of (^{14}C)α-aminoadipic acid and cysteinyl-valine into the tripeptide [12].

Some initial reports on the activity of AC or ACV synthesizing systems have been published although problems related to the determination of the enzyme activities remain unsolved since amino acid activation usually measured by the ATP-pyrophosphate exchange [13] can not be used with crude extracts. Adlington et al. [14] described the conversion of labeled α-aminoadipic acid, cysteine and valine into LLD-ACV by soluble extracts of C. acremonium. Although the incorporation of the label into the peptide was very small (0.1 to 0.4%), they also observed formation of the dipeptide LL-AC. Similar results were observed by López-Nieto [15] in our research group. Banko et al. [16] described an LL-AC synthetase activity in extracts of C. acremonium C10. The enzyme, which was stabilized by glycerol, carries out a lineal synthesis of LL-AC from its constituent amino acids for at least 6 h. The reaction is dependent on L-α-aminoadipate, L-cysteine, ATP and Mn^{2+} or Mg^{2+}. The optimal temperature is 25 °C to 30 °C. Ferrous ions, ascorbic acid and DTT, which are required for many of the subsequent reactions of cephalosporin biosynthesis do not affect the enzyme activity. The best time to harvest the mycelium so as to observe AC-synthetase activity is during the early phase of cephalosporin biosynthesis, e.g. 60 to 65 h of cultivation. Unfortunately, the activity of the enzyme is very low even in glycerol-protected preparations and this prevents purification.

Recently, production of ACV by glycerol stabilized cell-free extracts of C. acremonium has been reported [17]. The enzyme activity forms ACV from the individual amino acids L-α-aminoadipic acid, L-cysteine and L-valine in presence of ATP. Synthesis of ACV from the three individual component amino acids is far more rapid than tripeptide formation from LL-AC plus L-valine, what means that a single enzyme (ACV synthetase) must carry out the synthesis of LLD-ACV and this enzyme requires binding of the three amino acids for maximum activity [17]. The sterochemistry of the product formed was established as LLD-ACV. Therefore LD epimerization must occur during tripeptide formation since D-valine is not accepted as a substrate. The optimal concentration of ATP and Mg^{2+} were found to be 10 mM each, and the specific activity of ACV synthetase was about 20% higher in the presence of 10 mM Mg^{2+} than in the presence of 10 mM Mn^{2+}. The apparent K_m values of the ACV synthetase for the amino acids are, α-aminoadipic acid, 17 mM; cysteine 0,026 mM and valine 0,34 mM.

Fructose 1,6-diphosphate, glucose-1-phosphate and orthophosphate (which are known to inhibit also cephamycin biosynthetic enzymes) [18] were found to inhibit ACV synthetase. L-Glutamate was also inhibitory whereas L-methionine, glutathione, L-leucine, L-isoleucine and D-valine (all added at a concentration of 5 mM) had no effect. No compound having a positive effect upon the synthetase could be found [17]. Similar results were found using cell-free extracts of S. clavuligerus [19].

The ACV synthetase enzyme forms ACV analogues when provided with L-carboxymethylcysteine in place of L-α-aminoadipic acid or when provided with D-alloisoleucine or L-α-aminobutyric acid in place of L-valine. When glycine is added as a fourth amino acid no tetrapeptide is formed, although the corresponding tetrapeptide has been found in extracts of C. acremonium and P. persicinus [2, 20]. Very similar results have been found using extracts of S. clavuligerurs [19].

Further studies on the characterization of ACV synthetase are required before we can fully understand the molecular mechanisms involved in the formation of the tripeptide.

2.2 Isopenicillin N Synthase: Conversion of LLD-ACV to Isopenicillin N

The second enzyme of the common pathway, isopenicillin N synthase (cyclase) which hereinafter will be designated IPNS, has been purified from extracts of *P. chrysogenum* [9], *C. acremonium* [21, 22], *S. clavuligerus* [23] and *S. lactamdurans* [24]. The IPNS of the high penicillin producing strain *P. chrysogenum* AS-P-78 has been purified to near homogeneity by a combination of protamine sufate precipitation and ammonium sulfate fractionation, followed by dialysis, ion-exchange chromatography on DEAE-Sephacel and gel filtration on Sephacryl S-200 (Table 1). The estimated molecular weight obtained by gel filtration and polyacrylamide gel electrophoresis is 39,000

Table 1. Purification of isopenicillin N synthase of *P. chrysogenum*

	Total activity (pKats)	Protein (mg)	Specific activity (pKats mg^{-1})	Purification (fold)
1. Crude extract	1284	2140	0.6	1
2. Protamine sulfate	1190	1210	0.9	1.5
3. Ammonium sulfate	839	401	2.1	3.5
4. Sephacryl S-200	702	193	3.6	6
5. DEAE-Sephacel	533	49	11	18.3
6. Sephadex G-100	347	18	19.3	32.1
7. DEAE-HPLC	191	3.3	57.3	95.5

Data from Martín et al. [25]

Table 2. Characteristics of isopenicillin N synthases

Source of enzyme	Km^a	Molecular weight kDa	pI	Cofactor requirement	Inhibitors	Purification (folds)	Refs.
P. chrysogenum	0.13 mM	39	5.5^c	O_2, DTT, Fe^{2+} ascorbate	Co^{2+}, Mn^{2+} glutathion	95	9, 25)
C. acremonium	0.30 mM	40	5.0	O_2, DTT, Fe^{2+} ascorbate	Co^{2+}, Mn^{2+} α-aminoadipyl-cysteinyl-dihidrovaline	152	21)
S. lactamdurans	0.18 mM	26,5	6.5	O_2, DDT, Fe^{2+} ascorbate	Co^{2+}, Zn^{2+}, Mn^{2+} Glucose-6-phosphate	79	24)
S. clavuligerus	0.32 mM	33	ND	O_2, DTT, Fe^{2+} ascorbate	N-Ethylmaleimide	130	23)
A. nidulans	ND	37,4	ND^b	ND	ND	ND	42)

^{a} Km values for LLD α-aminoadipyl-cysteinyl-valine;
^{b} ND: not determined;
^{c} Barredo et al. (unpublished results)

$\pm$ 1,000, similar to the reported molecular weight of the isopenicillin N synthetase of *Cephalosporium acremonium* (Table 2) [9, 25]. The purified enzyme shows an apparent Km for ACV of 0.13 mM that is slightly lower than the reported value of the Km for the enzyme from *C. acremonium* (Table 2).

It requires dithiothreitol and is stimulated by ferrous ions and ascorbate, all of which are cofactors of oxygenases. Co^{2+} and Mn^{2+} completely inhibit the enzyme activity apparently due to competition with Fe^{2+}. The optimal temperature is 25 °C, similar to best temperature for penicillin production, and the optimal pH is 7.8. The reaction requires O_2 and is stimulated by increasing the dissolved oxygen concentration of the reaction mixture [9]. Cyclization, therefore, appears to take place by oxygen-mediated removal of four hydrogen atoms during formation of the β-lactam and thiazolidine rings.

The role of DTT in the cyclization reaction is dual — it protects the enzyme against inactivation by oxygen (itself required in the reaction) and it keeps the substrates in the adequate monomer form. The dimer ACV substrate is reduced to the monomer by DTT [26]. The monomer form of ACV appears to be the true substrate of the reaction. Moreover, DTT stabilizes the enzyme against inactivation during storage, indicating that the thiol groups of the enzyme are essential for the reaction (see below).

In cultures of *P. chrysogenum* IPNS is formed after the initial rapid growth phase (about 18 h) (Fig. 3) preceding the onset of penicillin biosynthesis (24 to 36 h). A high level of cyclase activity was present throughout the penicillin process [9] in contrast to the rapid disappearance of the cyclase in *C. acremonium* and *S. clavuligerus*. This result may account for the continued synthesis of penicillin for at least 120 h by the high penicillin producing strain AS-P-78. The time-course of IPNS is parallel to the pattern of ACV-forming activity [3] and to the overall penicillin "synthetase" activity [27].

P. chrysogenum strains with different abilities to produce penicillin differs greatly in IPNS activity. During strain improvement programs, mutations have resulted in

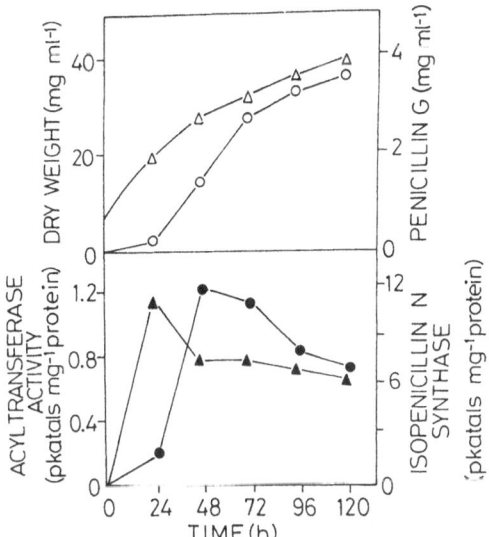

Fig. 3. Time course of growth (△), penicillin G formation (○), isopenicillin N synthase activity (▲) and 6-APA-acyltransferase activity (●) in batch cultures of *P. chrysogenum* grown in complex production medium (corn-steep based). Data from Ref. [9, 61]. Note the delay in the formation of 6-APA acyltransferase activity with respect to isopenicillin N synthase activity. Penicillin G is produced soon after enough 6-APA acyltransferase (the last enzyme of the pathway) is formed

drastic increases in IPNS activity. Thus, the high producing strain AS-P-78 contains a six-fold higher IPNS activity than that of strain Wis 54-1255, which correlates well with the five- to six-fold differences in antibiotic production between these two strains [9].

The IPNS of *C. acremonium* has been extensively studied [28, 29, 30, 31]. The cyclization reaction in extracts of *C. acremonium* is stimulated by Fe^{2+}, ascorbate and dithiothreitol, as occurred in *P. chrysogenum*, but not by ATP or α-ketoglutarate. The cyclase of *C. acremonium* has been highly purified and the molecular weight established as 40 to 42 kDa, similar to that reported for the cyclase of *Penicillium* (Table 2).

The purified IPNS of both *Penicillium* [9] and *Cephalosporium* [1, 22] gave one major band on analytical PAGE electrophoresis. The IPNS protein of *Cephalosporium* shows a pI of 5.0. Pang and co-workers [21] estimated that the rate of in vitro production of isopenicillin N by IPNS extracted from a given weight of mycelium was at least one order higher than the in vivo rate of synthesis of penicillin N and cephalosporin C together, by the same weight of mycelium at the time of harvesting. This observation suggests that under the experimental conditions used, the intracellular IPNS was not functioning under optimal conditions in the *C. acremonium* strain CO728, perhaps because the synthesis of ACV was rate limiting or cofactors were deficient, or both.

An involvement of thiol groups in the activity of IPNS of *C. acremonium* was also proposed based on the observation that blockage of at least one cysteinyl thiol inhibited IPNS activity [32]. This inhibition is in accordance with the proposed mechanisms of interaction of the IPNS with the ACV substrate by formation of a disulphide bond between ACV and a thiol group in the active site of the enzyme [33, 34]. The thiol groups may also play an important role in maintaining the proper conformation of IPNS. Baldwin and co-workers [35] observed that IPNS from *C. acremonium* could exist in both oxidized (inactive) and reduced (active) forms and suggested that an intramolecular disulphide bond was the difference between this two forms of the enzyme. These early observations on the requirement of thiol groups have been confirmed by a recent analysis of the role of cysteine residues in the IPNS of *C. acremonium* by site-directed mutagenesis (see below) [36].

2.2.1 Isopenicillin N Synthases of *Streptomyces*

The ability to synthesize β-lactam antibiotics, although restricted to a limited number of genera, is very frequent in members of the genus *Streptomyces*. At least 25 species of *Streptomyces* are known to produce different types of cephamycins [37].

The purification of IPNS from β-lactam producing *Streptomyces* has been hindered by the unavailability of high producing strains. However the IPNS of *S. clavuligerus* [26, 23] and *S. lactamdurans* [24] have been purified to near homogeneity using conventional columns and high performance liquid chromatography techniques. The IPNS of *S. clavuligerus* was purified 130-fold and that of *S. lactamdurans* 79-fold. The IPNS of *S. clavuligerus* has a molecular weight of 33 kDa and an apparent Km of 0.32 mM with respect to its substrate (ACV). The enzyme shows a sensitivity to thiol specific inhibitors with *N*-ethylmaleimide giving the strongest inhibitory effect [23] in agreement with previous observations made with the IPNS of *Penicillium* and *Cephalosporium*.

The IPNS of *Streptomyces lactamdurans* has also been purified to near homogeneity [24]. This enzyme has an absolute requirement for oxygen, Fe^{2+} and DTT and is greatly stimulated by ascorbic acid, i.e. it behaves as an oxygenase although it does not require 2-oxoglutarate. The enzyme activity is strongly inhibited by Co^{2+}, Zn^{2+} and Mn^{2+} probably due to competition with the Fe^{2+} requirement. Optimal pH and temperature of the enzyme are 7.0 and 25 °C. The pI of the cyclase of *S. lactamdurans* is 6.55 whereas the IPNS of *C. acremonium* has a pI of 5.05. The purified enzyme showed an apparent Km for ACV of 0.18 mM, similar to the Km of the cyclase of *P. chrysogenum* (0.13 mM) and slightly lower than the reported value for *C. acremonium* (0.3 mM) and *S. clavuligerus* (0.32 mM).

The molecular weight of the IPNS of *S. lactamdurans* (26,5 kDa) is clearly smaller than that of the cyclases of *P. chrysogenum* (39 kDa), *C. acremonium* (38,416) and *S. clavuligerus* (33 kDa). The enzyme seems to be a monomer since the molecular weight of the natural (non-denatured) form calculated by gel filtration is identical to that of the SDS-denatured protein [24].

To sum up, the characteristics of the four cyclases purified so far are summarized in Table 2. Some of the differences in the properties and kinetic parameters of the enzymes, e.g. molecular weight, reflect evolutionary changes that the enzymes have suffered in different organisms. Other differences (e.g. in specific activity) may be due to the different methods of cultivation used to grow mycelium or to differences in the preparation of cell-free extracts and purification of the enzymes.

2.2.2 Cloning and Molecular Characterization of the Isopenicillin N Synthase Genes

Cloning of the IPNS gene from *C. acremonium* was acomplished by purifying IPNS from *C. acremonium* ATCC11550, followed by determining the first 23 amino terminal amino acids. A set of synthetic oligonucleotides based on the amino terminal amino acid sequence was prepared and a cosmid genome library was probed with the mixed oligonucleotides. A subsequent determination of the N-terminal amino acid sequence of the IPNS from *C. acremonium* CO728 confirmed the original data except for the amino acid in position 21 [35]. The IPNS gene was trimmed down and sequenched. The open reading frame encodes a polypeptide of Mr 38,416 which has a predicted amino acid sequence that matched the experimentally determined sequence (Fig. 4). When this open reading frame was inserted into an *E. coli* expression vector and transformed into *E. coli*, the recombinant strain produced a new protein co-migrating with authenthic IPNS as the mayor protein of the cell (about 20% of the cell protein) [38]. The predicted amino acid sequence of the IPNS protein begins with methionine and glycine residues which are not found in the protein isolated from *Cephalosporium* cells, which suggests that these residues are cleaved post-translationally [39].

The IPNS has been isolated from *Escherichia coli* transformed with the cloned gene from *C. acremonium* and purified to homogeneity [40]. The protein synthesized in *E. coli* has undergone a slightly different *N*-terminal processing to that observed in the fungal protein. In *C. acremonium* mature IPNS had lost the terminal methionine and glycine residues whereas in the *E. coli* IPNS the terminal methionine residue is removed but the second residue, glycine, is not. The different processing observed in *E. coli* has no apparent consequences on the biochemical properties of the enzyme. The Km and kinetics for the conversion of LLD-ACV to isopenicillin are identical. Recombinant IPNS converted analogue substrates into unusual beta-lactam antibiotics in exactly the same way as the fungal protein [40].

```
                                                                                                57
S.c ATG --- CCA GTT CTG ATG CCG TCG GCC CAC GTT CCG ACC ATC GAC ATC TCG CCG CTG
C.a ATG GGT TCC GTT CCA GTT CCA GTG GCC AAC GTC CCC CGA ATC GAT GTC TCG CCC CTA
P.c ATG GGT TCC ACC --- --- CCC AAG GCC AAT GTC CCC AAG GTG TCG CCC CTG
A.n ATG GGT TCA GTC --- --- AGC AAA GTC AAT GTT CCA AAG ATC GAT GTT TCT CCC CTG

                                                                                                114
    TTC GGA ACC GAC GCC GCC GCG AAG AAG CGC GTC GCC GAG GAG ATA CAC GGG GCC TGC
    TTC GGC GAT GAC AAG GAG AAG AAG CTC GAG GTA GTC CGC GCC ATC GAC GCC GGA TCG
    TTC GGC GAC AAT ATG GAG GAG AAG ATG AAG GTT GCC CGC GCG ATT GAC GCT GCC TCG
    TTT GGA GAC GAC CAA GCA GCC AAA ATC AGA GTA GCC CAG CAA ATC GAC GCC GCT TCG

                                                                                                171
    CGC GGC TCG GG_ TTC TTC TAC GCC ACG AAC CAC GGC GTG GAC GTC CAG CAG CTC CAG
    CGC GAC ACA GGC TTC TTT TAC GCC GTG AAC CAC GGT GTC GAC CTG CGG CTC TCG
    CGC GAC ACC GGC TTC TTC TAC GCG GTC AAC CAC GGT GTG GAT GTG AAG CGA CTC TCG
    CGT GAT ACT GGG TTT TTC TAC GCC GTC AAC CAT GGG ATC AAT GTG CAG CGA CTC TCG

                                                                                                228
    GAC GTG GTG AAC GAG TTC CAC GGC GCC ATG ACC GAC CAG GAG AAG CAC GAC CTG GCG
    CGC GAG AAC AAA TTC CAC AGC ATC GAC GAC GAG AAG CAC GAC CTG GCC
    AAC AAG ACC AGG GAG TTC CAC TTT TCT ATC ACA GAC GAA GAG AAG TGG GAC CTC GCC
    CAG AAG ACC AAG GAG TTT CAT ATG TCT ATC ACA CCT GAG GAA AAG TGG GAC CTT GCG

                                                                                                285
    ATC CAC GCG TAC AAC CCG GAC AAC CCG --- CAT GTG CGC AAC GGC TAC TAC AAG GCG
    ATC CGG GCC TAC AAC AAG GAG CAC GAG TCC CAG ATC CGG GCG GGC TAC TAC CCG CCG
    ATT CGC GCC TAC AAC AAG GAG CAC CAG GAC CAG ATC CGT GCC GGA TAC TAC CTG TGC
    ATT CGT GCC TAC AAC AAA GAG CAC CAG GAC CAA GTC CGT GCC GGA TAC TAC CTG TCC

                                                                                                345
    GTC CCG ┌GGC AGG AAG GCC GTC GAG TCC TTC TGT TAC┐ CTC AAC CCG GAC TTC GGC GAG
    ATC CCG │GGC AAG AAG GCG GTC GAA TCG TTC TGC TAC│ CTG AAC CCC TCC TTC AGC CCA
    ATT CCG │GAG AAA AAG GCC GTG GAC TCC TTC TGC TAC│ CTG AAC CCC AAC TTC ACG CCC
    ATC CCC └GGG AAA AAG GCA GTC GAG TCC TTC TGC TAC┘ CTT AAC CCG AAC TTC ACG CCC

                                                                                                399
    GAC CAC CCG ATG ATC GCC GCG GGG ACG CCG ATG CAC GAG GTG AAC CTC TGG CCC GAC
    GAC CAC CCG CGA ATC AAG GAG CCC ACC CCT ATG CAC GAG GTC AAC GTC TGG CCG GAC
    GAC CAC CCT CTC ATC CAG TCG ACT CCC ACT CAC GAG GTC AAC GTG TGG CCG GAC
    GAT CAT CCC CGT ATC CAG GCC AAG ACT CCC ACT CAC GAG GTA AAC GTG TGG CCA GAC

                                                                                                456
    GAG GAG CGG CAC CCG CGC TTC CGG CCG TTC TGC GAG GGC TAC TAC CGG CAG ATG CTG
    GAC GCG AAG CAC CCG GGG TTC GCC TTC GCC GAG AAG TAC TAC TGG GAC GTC TTC
    GAG AAG AAG CAT CCG GGC TTC CGC GAG TTC GCC GAG CAA TAC TAC TGG GAT GTG TTC
    GAG ACC AAG CAC CCT GGT TTC CAA GAC TTT GCC GAG CAG TAT TAC TGG GAT GTT TTC

                                                                                                513
    AAG CTC TCC ACC GTG CTC ATG CGG GGG CTG GCG CTG GCG CTC GGG AGG CCG GAG CAC
    GGC CTG TCC TCC GCG GTG CTC CGC GGG TAC GCT CTC GCC CTA GGT CGC GAC GAG GAC
    GGG CTC TCG TCT GCC TTG CTG CGA GGC TAT GCT CTG GCG CTG GGC AAG GAG GAG GAC
    GGT CTC TCT TCT GCA CTG CTC AAG GGC TAC GCC TTG GCA TTA GGC AAA GAG GAG AAT

                                                                                                570
    TTC TTC GAC GCG GCG CTC GCC GAG CAG GAC TCC CTG TCG TCC GTC TCG CTG ATC CGC
    TTC TTC ACC CGC CAC TCC CGC CGT GAC ACG CTC TCG TCG GTC CTC ATC CGC
    TTC TTT AGC CGC CAC TTC AAG AAG GAA GAC GCG CTC TCC TCG GTT GTT CTG ATT CGT
    TTC TTC GCT CGC CAC TTC AAG CCA GAC TAT CTT GCC TCG GTT GTG CTG ATC CGT

                                                                                                627
    TAC CCG TAT CTG GAG GAC TAC CCG --- CCG --- GTG AAG ACG GGT CCC GAC GGC CAG
    TAC CCG CAC CTC GAC CCG TAC CCG GCG CCG --- ATC AAG ACG GCG GAC GAC GGC ACC
    TAC CCG TAC AAC CCC ATC CCA CCT GCC GCC ATT AAG ACG GCG GAG GAC GGC ACC
    TAC CCT TAC CTG GAT CCC TAC CCC GAG GCT GCT ATC AAG ACG GCG GCC GAC GGC ACC

                                                                                                684
    CTC CTG AGC TTC GAG GAC CAT CTG GAC GTC TCG ATG ATC ACC GTG CTC TTC CAG ACC
    AAG CTC AGC TTC GAG TGG CAC GAG GAC GTG TCC CTC ATC ACG GTG TTG TAC CAG TCC
    AAA TTG AGT TTC GAA TGG CAT GAG GAC GTG TCG CTC ATT ACC GTG CTG TAC CAG TCA
    AAA CTG AGT TTT GAG TGG CAT GAG GAT GTA TCC CTA ATC ACT GTG CTT TAC CAG TCC

                                                                                                741
    CAG GTG CAG AAC CTC CAG GTG GAG ACG GTC GAC GGC TGG CGG GAC ATC CCG ACG TCG
    GAC GTG CAG AAT CTC CAG GTG AAG ACC GAC GGC TGG CAG GAC ATC CAG GCT GAC
    GAC GTG GCG AAC CTG CAG GTG GAG ATG CCC CAG GGT TAC CTC GAT ATC GAG GCG GAC
    AAC GTG CAG AAC CTG CAG GTA GAA ACT GCT GCC GGG TAT CAA GAT ATC GAG GCC GAC

                                                                                                798
    GAG AAC GAC TTC CTG GTC AAC TGC GGT ACC TAC ATG GCG CAT GTC ACG AAC GAC TAC
    GAC ACG GGC CTC ATC AAC TGC GGC AGC TAC ATC CCC CAT ATC ACC GAC GAC TAC
    GAC AAC GCC TAC CTG GTC AAT TGC GGC AGC TAC ATG GCA CAC ATC ACC AAC AAC TAC
    GAT ACT GGC TAC TTG ATC AAC TGC GGC AGT TAC ATG GCA CAT CTT ACA AAC AAC TAC

                                                                                                855
    TTC GCG GCG CCC AAC CAC CGG GTG AAG TTC GTG AAC GCG GAG CGG CTG TCC CTG CCG
    TAC GCG GCC CCC ATC CAC CGG GTC AAA TGG GTC AAG GAG GAG CGC CAG TCA CTC CCC
    TAC CCC GCT CCC ATC CAC CGG GTC AAG TGG GTG AAC GAG GAG CGC CAA TCC CTC CCG
    TAT AAA GCG CCC ATC CAT CGG GTG AAA TGG GTT AAT GCA GAG CGC CAG TCC CTG CCA

                                                                                                912
    TTC TTC CTC AAC GGC GGG GAC GAG GCG GTC ATC AAC GCG CCG TTC GTG CCG --- --- ---
    TTC TTC GTC AAC CTG GGC TGG GAG GAC ACC ATC CAG CCG TGG GAC CCC GCG ACC GCC
    TTC TTC GTC AAT CTG GGA TTT AAT GAT ACC GTC CAC CCG TGG GAT CCT --- AGC ---
    TTC TTC GTC AAC CTG GGA TAC GAC GAC GTG ATT GAT CCA TTT CCC --- --- ---

                                                                                                969
    --- GAG GGC --- GCG AGC GAG GAG GTG AGG --- --- AAC --- GAG GCC CTG TCC TAC
    AAG GAT GGG --- GCC AAG GAT GCC GCC AAG GAC --- AAG --- CCG GCC ATC TCC TAC
    AAG GAA GAC --- GCC AAG --- --- --- ACC GAT CAG CGG --- CCA --- ATC TCG TAC
    CGA GAA CCC AAT GGC AAG --- --- --- TCT GAT --- CGG GAG CCA --- CTC TCC TAC

                                                                                                1023
    GGG GAC TAC CTC CAG CAC GGG CTG CGG GCG CTG ATC GTC AAG AAC GGC CAG ACC   *
    GGA GAC TAT CTG CAG GGG GGA CTG CGG GGC TTG ATC AAC AAG AAT GGT CAG ACC   *
    GGC GAC TAT CTG CAG AAC GGA TTA GTT AGT CTA ATC AAC AAG AAC GGC CAG ACA   *
    GGC GAC TAT TTG CAA AAC GGG CTG GTG AGT TTG ATC AAC AAG AAC GGC CAG ACC   *
```

The IPNS gene of *P. chrysogenum* has also been cloned in two different laboratories. We cloned *Sau3A*-digested fragments of total DNA of *P. chrysogenum* AS-P-78 into the *Bam*HI site in EMBL-3 arms, a λ phage derived vector (Barredo et al. [40a]). The same strategy was followed by Carr et al. [41] to clone the ÌPNS gene of a different strain of *P. chrysogenum* 23X-80-269-37-2. The library was screened using a heterologous hybridization probe based on the nucleotide sequence of the amino terminal end of the IPNS gene of *C. acremonium*. *E. coli* cells transformed with the IPNS gene contained IPNS activity whereas untransformed cells were devoid of it. Results of both laboratories indicate that the cloned open reading frame (Fig. 4) encodes a polypeptide with a molecular weight of 37.9 kDa which agrees with the reported value of 39 ± 1 kDa for the purified protein from *P. chrysogenum* [9]. However there is one amino acid which is different in the IPNS proteins from strain AS-P-78 and the high producer 23X-80-269-37-2.

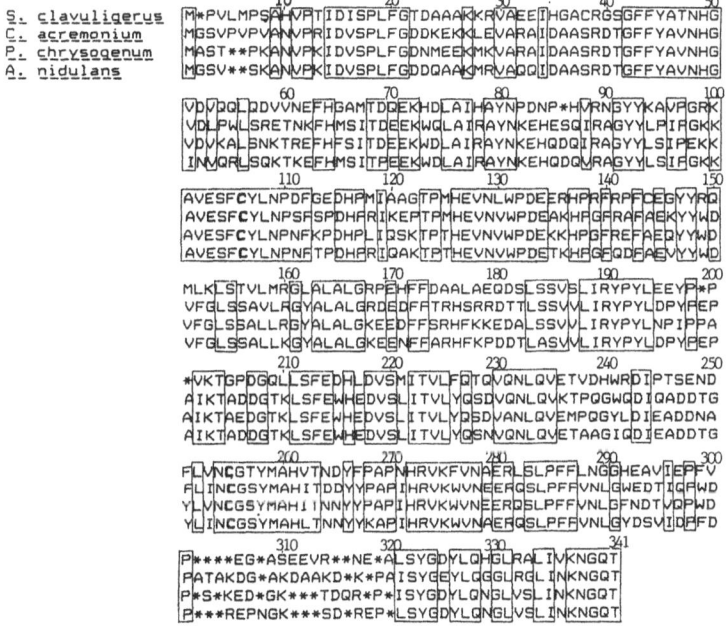

Fig. 5. Amino acid sequences of the isopenicillin N synthases of *S. clavuligerus*, *C. acremonium*, *P. chrysogenum* and *A. nidulans*. Some gaps (*) have been inserted to obtain maximal homology. The two cysteines (amino acids 106 and 255 of the *C. acremonium* sequence) are in boldface letters. Identical or functionally conserved amino acids in the four polypeptides are boxed. (Data from Refs. [38,41, 42,42a])

◄ **Fig. 4.** Nucleotide sequences of the isopenicillin synthases of *S. clavuligerus*, *C. acremonium*, *P. chrysogenum* and *A. nidulans*. Some gaps (———) have been introduced in the nucleotide sequences of the different genes to obtain maximal nucleotide homology. (Data from Refs. [38,41,42,42a]). The boxed sequence has a high degree of similarity with the boxed region in the nucleotide sequence of the expandase gene of *C. acremonium*. S. c., *Streptomyces clavuligerus*; C. a., *Cephalosporium acremonium*; P. c., *Penicillium chrysogenum* and A. n., *Aspergillus nidulans*

More recently the IPNS gene of *Aspergillus nidulans* has been identified by heterologous hybridization with a DNA probe corresponding to the *C. acremonium* IPNS [42].
The open reading frame encodes a 331 amino acid polypeptide with extensive homology
with the genes of other β-lactam producing fungi and a molecular weight of 37,480.
The gene product has been overexpressed in *E. coli* giving IPNS activity.

The isopenicillin N synthase gene of *S. clavuligerus* has also been cloned and
resembles those of *P. chrysogenum* and *C. acremonium* [42a].

No evidence for introns was found in the open reading frames of the IPNS genes
from either *Penicillium* or *Cephalosporium*. This fact together with the high G+C
content (about 63%) of the open reading frames, suggest the IPNS genes originate
from the DNA of β-lactam producing *Streptomyces*. Attempts to draw conclusions
regarding important residues in both IPNS proteins were initially hampered by the
extensive homology [41]. Many residues were conserved (the genes of *C. acremonium*,
P. chrysogenum, *A. nidulans* and *S. clavuligerus* match at least 8 amino acid residues
in a row in 12 separate regions) and it was difficult to assign significance to conserved
regions (see boxed areas in Fig. 5). More significative differences are observed in the
gene of *S. clavuligerus*.

All IPNS proteins contain two cysteine residues (numbered cys-255 and cys-106
in the *C. acremonium* enzyme) nested on highly conserved regions which extent
about 5 amino acids on each side of the cysteines. Also both cysteines have a histidine
residue within 5 to 10 amino acids downstream from the cysteine [41]. One or both
of these histidines may be involved in iron binding since histidine is known to be
commonly involved in iron binding [43].

2.2.3 Site Directed Mutagenesis of the IPNS Gene

The role of the two cysteine residues in the activity of the *C. acremonium* enzyme has
been examined by using site directed in vitro mutagenesis in order to change the
cysteines into serine residues [36]. Serine was chosen because the shape of the molecule
is very similar to that of cysteine but instead of a sulfur atom serine contains an oxygen
atom. Mutation of the cys-255 reduces specific activity by about 50% whereas mutation of cys-106 or mutation of both cys-106 or cys-255 reduces specific activity by
about 97%. In fact IPNS is able to catalyze cyclization of ACV to isopenicillin N
even when both sulphur atoms are replaced by oxygen atoms although the specific
activity is reduced by a factor of 30. Changing cys-106 to a serine residue is associated
with a 5-fold increase in the apparent Km for ACV whereas changing cys-255 to a
serine residue has virtually no effect on the Km. Based on these results, Samson
et al. [36] have concluded that the cysteine residues are important but not necessary for
IPNS activity. These authors proposed four possibilities to explain the role of the
sulfhydryls in IPNS activity:

1) the sulfhydryl(s) may simply play a structural role, stabilizing the proper conformation of the protein;
2) the sulfhydryl(s) may carry out a catalytic role, directly involved in electronic
   interaction with the substrate in the active site;
3) the sulfhydryl(s) may be indirectly involved in catalysis, perhaps by binding

and/or stabilizing the substrate or a cofactor such as iron or molecular oxygen; or

4) the role of the sulfhydryls may be some combination of the above.

The most likely role for the sulphydryls in IPNS is the involvement in binding and/or stabilizing the substrate or cofactors such as iron or molecular oxygen. This would explain why covalent modification of the sulphydryl groups dramatically reduce activity of the protein, since addition of an alkyl group may sterically exclude the iron from the active site, whereas substitution of an oxygen for the sulphur would still allow the iron to reach partially the active site [36]. An additional possibility is that the sulphydryl groups are involved in substrate binding but not directly implicated in catalysis. This possibility is supported by the observation that the Km of the protein with the ser-106 mutation is about 5-fold higher than for proteins with a cysteine at position 106. The cysteine residue at position 106 therefore appears to be involved in substrate binding either directly or indirectly [36].

2.2.4 Enzymatic Synthesis of New Penicillins Using the Isopenicillin N Synthases

The availability of semipurified or pure preparations of IPNS from different sources makes it possible to test a variety of analogues of the natural substrate ACV so as to try to produce new penicillins. This strategy has been somewhat successful despite the narrow substrate specificity of the isopenicillin N synthases [44].

Early studies on substrate specificity of the IPNS from *C. acremonium* indicated that IPNS does not tolerate large changes in the molecule of ACV [31, 45, 26] although, to some degree, it accepts the removal of the terminal amino group of α-aminoadipic acid [46] or the substitution of a CH_2 by a sulfur atom [47]. In fact, Bahadur et al. [31] described the synthesis of a new β-lactam antibiotic by cyclization of a tripeptide in which the valine component of ACV had been substituted by isoleucine. Moreover, δ-(L-carboxymethylcysteinyl)-L-cysteinyl-D-valine, a sulfur analogue of the LLD-ACV is converted to a sulfur analogue of isopenicillin N [47,48,49]. A new antibacterial penicillin containing a 2-α-methoxy-group was obtained from the tripeptide δ-(L-α-

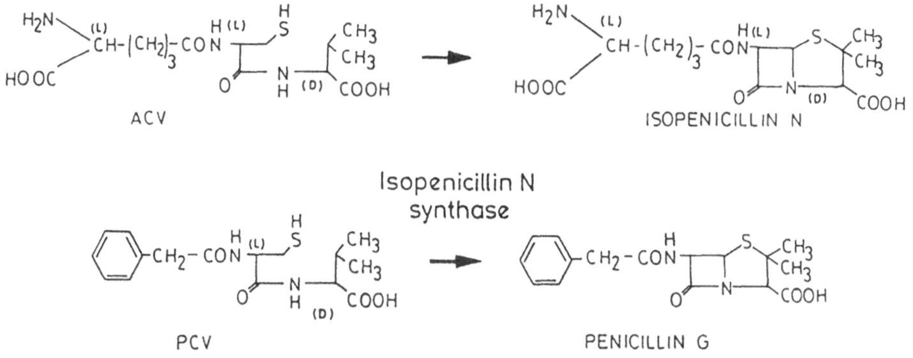

Fig. 6. Comparative cyclization of LLD-ACV and phenylacetyl-L-cysteinyl-D-valine (PCV) by isopenicillin N synthases of *A. chrysogenum*, *P. chrysogenum*, *S. lactamdurans* and *S. clavuligerus* (see text)

aminoadipyl)-L-cysteinyl-D-(O-methylallothreonine) with IPNS obtained from *C. acremonium* [50].

The IPNS does not tolerate any shortening of the side chain even by one carbon atom [49] and therefore, it was proposed that a six carbon or equivalent chain terminating in a carboxyl group, was the minimum requirement for penicillin synthesis by the IPNS [51].

It was, therefore, surprising to find that a direct enzymatic synthesis of penicillin G could be carried out by in vitro cyclization of the linear tripeptide-like molecule phenylacetyl-L-cysteynil-D-valine (PCV) (Fig. 6) using purified cyclases of *P. chrysogenum* and *A. chrysogenum* [52]. We also observed a similar cyclization using extracts of *S. lactamdurans* [53].

Table 3. Peptides substrates for isopenicillin N synthases

Substrate analogs	Source of enzyme	Conversion efficiency or Km	Refs.
L-(α-Aminoadipyl)-L-cysteinyl-D-valine	*P. chrysogenum*	0.13 mM	9)
(ACV)	*C. acremonium*	0.3 mM	70)
	S. clavuligerus	0.32 mM	23)
	S. lactamdurans	0.18 mM	24)
Phenylacetyl-L-cysteinyl-D-valine	*P. chrysogenum*	1.7 mM	52)
(PCV)	*C. acremonium*	1.8 mM, 0.12%	52, 132)
	S. lactamdurans	3.6 mM	53)
	S. clavuligerus	+	54, 55)
Phenoxyacetyl-L-cysteinyl-D-valine	*C. acremonium*	0.8 mM, 0.04%	56)
2,5-Cyclohexadienoyl-L-cysteinyl-D-valine	*C. acremonium*	+	
m-Acetyl-phenylacetyl-L-cysteinyl-D-valine	*C. acremonium*	0.11%	56)
m-Acetyl-phenoxyacetyl-L-cysteinyl-D-valine	*C. acremonium*	0.17%	134)
p-Acetyl-phenylacetyl-L-cysteinyl-D-valine	*C. acremonium*	0.02%	134)
p-Acetyl-phenoxyacetyl-L-cysteinyl-D-valine	*C. acremonium*	0.03%	134)
∂(D-α-Aminoadipyl)-L-cysteinyl-D-valine	*C. acremonium*	+	135)
Adipyl-L-cysteinyl-D-valine	*C. acremonium*	+	47)
∂(L-α-Aminoadipyl)-L-cysteinyl-D-(O-methyl-D-allothreonine)	*C. acremonium*	+	50)
∂(L-α-Aminoadipyl)L-cysteinyl-D-(O-methyl-D-threonine)	*C. acremonium*	+	50)
∂(L-α-Aminoadipyl)-L-cysteinyl-D-isoleucine	*C. acremonium*	+	50)
D-Carboxymethylcysteine-cysteinyl-valine	*C. acremonium*	+	47)
L-Carboxymethylcysteine-cysteinyl-valine	*C. acremonium*	+	47)
∂(L-Carboxymethylcysteinyl)-L-cysteinyl-D-valine	*C. acremonium*	+	47)
m-Carboxyphenylacetyl-L-cysteinyl-D-valine	*C. acremonium*	0.8 mM	133)
L-Glutamyl-L-cysteinyl-D-valine	*C. acremonium*	No	57)
L-Aspartyl-L-cysteinyl-D-valine	*C. acremonium*	No	57)
N-Acetyl-L-α-aminoadipyl-L-cysteinyl-D-valine	*C. acremonium*	No	57)
Glycyl-L-α-aminoadipyl-L-cysteinyl-D-valine	*C. acremonium*	No	57)
L-α-Aminoadipyl-L-cysteinyl-D-isoleucine	*C. acremonium*	+	57)
L-α-Aminoadipyl-L-cysteinyl-D-ethyl,propyl-glycine	*C. acremonium*	+	57)
∂(L-α-Aminoadipyl)-L-cysteinyl-glycine (ACG)		Inhibitor	137)

+ Positive cyclization without determination of the Km

The reaction product shows an antibiotic spectrum similar to that of penicillin G; it is sensitive to β-lactamases and penicillin acylases, and has been identified by chromatography, electrophoresis and HPLC as penicillin G. Similar results were later observed with the cyclase of *S. clavuligerus* [54, 55].

In all cases, the efficiency of cyclization of PCV was very small. Baldwin et al. [56] confirmed that peptides containing aromatic moieties (such as phenoxyacetyl-cysteinyl-valine and phenylacetyl-cysteinyl-valine) or related unsaturated molecules (e.g. 2,5-cyclohexadienoyl-L-cysteinyl-D-valine) in place of the α-aminoadipyl side chain can in fact be cyclized to the corresponding penicillins albeit at very low rates. (Table 3).

The low efficiency of cyclization of PCV and related compounds may explain the apparent discrepancy between our results and the minimal structural requirements of the N-acyl group proposed by Baldwin et al. [51]. The side chain of PCV has more than 6 carbon atoms and does not terminate in a carboxyl group which explains the difference in affinity of the enzyme for ACV and PCV.

Confirmation that unusual β-lactams are generated by competing reactions all catalyzed by IPNS was strongly supported by the incubation of LLD-aminoadipyl-cysteinyl-allylglycine with *C. acremonium* IPNS obtained from the cloned gene in *E. coli* [40]. A mixture of six β-lactam products are produced by the *E. coli* enzyme as well as by the fungal enzyme.

The possibility of using cyclases from different organisms to obtain directly penicillin G or other antibiotics by cyclization of PCV or ACV analogues may have an important industrial relevance. Penicillin G or other β-lactam antibiotics may be produced in the future by cyclization of peptides using soluble or immobilized cyclases [57]. To achieve this goal it will be necessary to alter the substrate specificity and/or the affinity of cyclases for the different substrates by site directed in vitro mutagenesis of the cloned IPNS genes.

2.2.5 Amplification of the IPNS Activity in *Cephalosporium* Producing Strains

The IPNS gene has been amplified by transforming *C. acremonium* with vectors constructed by splicing the promoter region of the IPNS gene to the open reading frame of the bacterial hygromycin phosphotransferase gene [58]. These vectors have been used to insert multiple copies of the IPNS gene in a non-producer mutant of *C. acremonium* N_2 and in the wild type *C. acremonium* strain. Some of the transformants of the N_2 mutants regained the isopenicillin N synthase activity. When the cephalosporin C titers of 100 transformants of the wild type were compared to the titers of 100 natural selectants, some transformants exhibited titers significantly higher than the wild type. However none of the transformants had titers superior to the best titers obtained with natural selectants of the wild type [39]. This suggests that the IPNS may not be a rate limiting enzyme in the biosynthesis of cephalosporin C in this strain of *C. acremonium*. However it might be necessary to study a large number of transformants in order to find a strain that carries mutiple inserts of the IPNS gene integrated in such a way that they are not deletorius for the genome of the transformed strain [39, 59].

3 Late Enzymes of the Penicillin Pathway

3.1 Conversion of Isopenicillin N to Penicillin G or Other Hydrophobic Penicillins

Biosynthesis of aromatic penicillins proceeds necessarily through the intermediate isopenicillin N that contains an α-aminoadipyl side chain. In the last reaction of the penicillin biosynthetic pathway the α-aminoadipyl side chain is exchanged for phenyl-acetic acid previously activated in the form of phenylacetyl-CoA. This reaction is carried out by an acyltransferase complex found in crude extracts of *P. chrysogenum* [60, 61]. In the absence of side chain precursors, *P. chrysogenum* produces significative amounts of 6-aminopenicillanic acid (6-APA) [62, 63, 64] and small amounts of several different penicillins, e.g. benzylpenicillin, *p*-hydroxybenzylpenicillin, 2-pentenyl-penicillin, *N*-amylpenicillin, *N*-ethylpenicillin and isopenicillin N [10]. These are formed from monosubstituted acetic acids that exist at low concentration in the cell. Addition

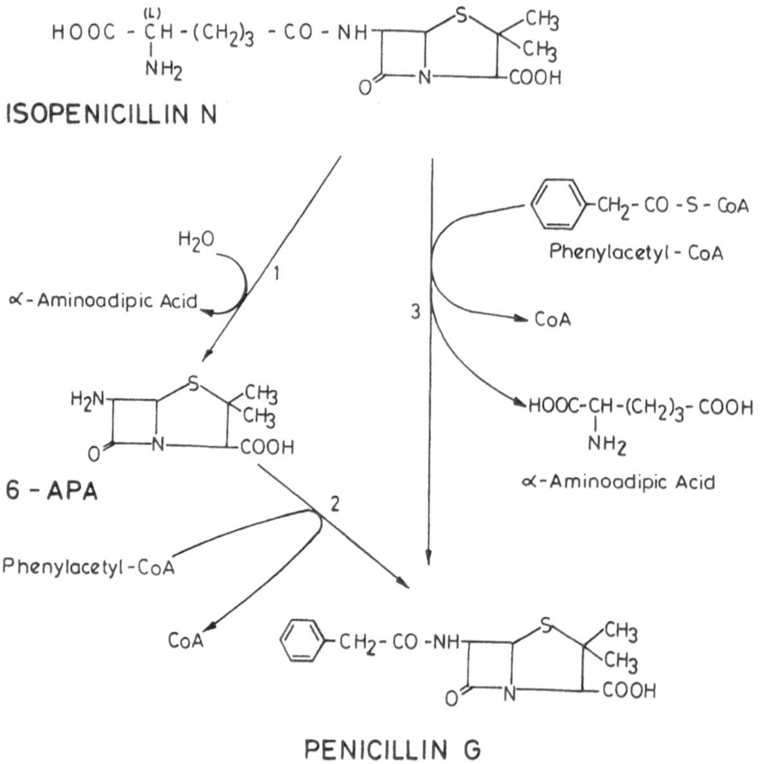

Fig. 7. Two-step versus direct conversion of isopenicillin N into penicillin G. In the two-step con-version the α-aminoadipyl side chain of isopenicillin N is cleaved by isopenicillin N amidase (6-APA forming) (1) and 6-APA is later acylated by the acyl-CoA:6-APA acyltransferase (2). Pure preparations of acyl CoA:6-APA acyltransferase contain also isopenicillin N acyltransferase (3) which converts directly isopenicillin N into penicillin G (see text)

of phenylacetic acid to cultures of *P. chrysogenum* produces an increase of penicillin G so that less 6-APA and other penicillins are formed.

The role of 6-APA in the de novo biosynthesis of penicillin is unclear. Demain [6] proposed that the terminal reaction of benzylpenicillin synthesis is most likely an exchange of the L-α-aminoadipic acid side chain of isopenicillin N for phenylacetic acid activated in the form of phenylacetyl-CoA. However, a two step reaction involving 6-APA as an intermediate is also possible (Fig. 7). Reports on the presence of (a) an acylase (or amidolyase) in *P. chrysogenum* which removes the side chain of benzyl-penicillin to yield 6-APA and (b) an acyl-CoA:6-APA acyltransferase which catalyzes a direct exchange of side chains between isopenicillin N and solvent soluble penicillins (or between such penicillin and 6-APA) are contradictory [65, 66, 67].

Later, Abraham and co-workers [68] found that an enzyme (or enzymes) in crude extracts of *P. chrysogenum* catalyzes the formation of benzylpenicillin from iso-penicillin N as well as from 6-APA in the presence of phenylacetyl-CoA. Attempts to solve the dilemma of the numerous activities of the "acyltransferase complex" were hampered by the extreme instability of the enzyme(s) in crude extracts. Only recently has the acyl-CoA:6-APA acyl transferase of *P. chrysogenum* AS-P-78 been purified to homogeneity (258-fold) as concluded by SDS polyacrylamide gel electro-phoresis and isoelectric focusing (Table 4 and Fig. 8) [61]. The enzyme is a monomer with a molecular weight of 30 kDa and a pI of about 5.5. The molecular weight of the acyltransferase is slightly lower than the molecular weight of the IPNS of *P. chrysogenum* (39 kDa) [9, 41] and *C. acremonium* (Mr 38416) [38]. This enzyme was found in two different penicillin producing strains of *P. chrysogenum* (Wisconsin 54-1255 and AS-P-78) and in *A. nidulans* [Alvarez, E., unpublished] but was absent in 4 *npe* (blocked in penicillin biosynthesis) mutants isolated in our laboratory, suggesting that the acyl-CoA:6-APA acyltransferase is undoubtedly involved in penicillin bio-synthesis [25]. The optimal acyltransferase temperature (25 °C) agrees with the tem-perature at which the production of penicillin is optimal. In liquid cultures, onset of 6-APA acyltransferase occurs after formation of isopenicillin N synthase which is synthesized following growth of *P. chrysogenum* (Fig. 3) [61]. The sequential formation of the biosynthetic enzymes of cephalosporins [69, 70] and other secondary metabolites [71]

Table 4. Purification of the acyl-CoA:6-APA acyltransferase of *Penicillium chrysogenum*

Purification steps	Acyl transferase activity (pkatals)	Protein (mg)	Specific activity (pkatals mg^{-1} of protein) 6-APA[a]	Isopenicillin N[a]	Recovery (%)	Purification (fold)
Crude extract	3888	3240	1.2	0.2	100	1.0
Protamine sulfate supernatant	3630	1160	3.1	0.6	93	2.6
Ammonium sulfate (40–55%) precipitate	2510	410	6.1	1.1	64	5.1
Phenyl-sepharose	1060	14	75.7	16.3	27	63.1
DEAE-sephacel	740	3.5	211.4	38.2	19	167.0
Sephadex G-75	310	1	310	55.3	8	258.3

[a] 6-Aminopenicillanic acid (6-APA) or isopenicillin N are used as substrates by the enzyme complex.
Data from Alvarez et al. [61]

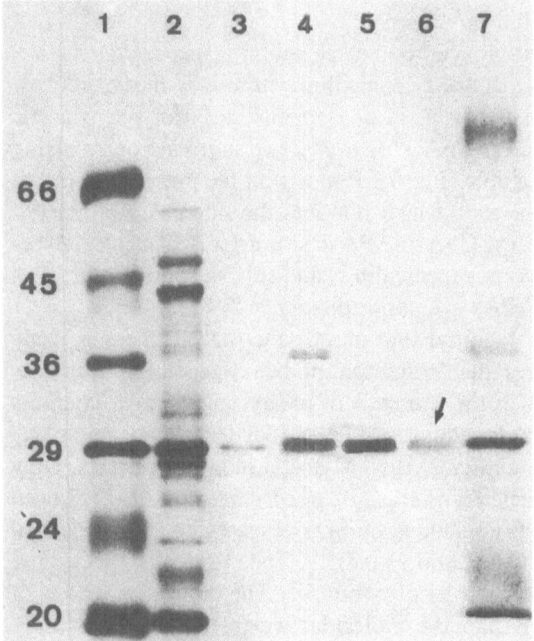

Fig. 8. SDS-PAGE of fractions from different steps of purification of acyl-CoA : 6-APA acyltransferase. Lanes 1 and 7, molecular weight standards (in thousands); lane 2, eluate from the Phenylsepharose CL-4B column; lanes 3 and 4, fractions showing peak enzyme activity from the DEAE-Sephacel column; lanes 5 and 6, two different amounts (approximately 16 and 8 μg, respectively) of pure acyltranferase after Sephadex G-75 gel filtration. The acyltransferase band is indicated by an arrow (from Ref. [61])

appears to be a general phenomenon in fungi. The specific activity of the acyl-CoA : 6-APA acyltransferase declines at the same time as does the penicillin production rate.

The acyltransferase showed a high affinity and specificity for 6-APA (Km 9.3 μM). It did not accept 7-ACA, benzylpenicillin, cephalosporin C or isocephalosporin C. An isopenicilin N acyltransferase observed in crude extracts of *P. chrysogenum* copurify with the 6-APA acyltransferase. However, the affinity of the enzyme for isopenicillin N (and also for penicillin N) as substrates is smaller than for 6-APA [61].

The pure acyltransferase does not show penicillin hydrolyzing activity (penicillin acylase) which splits penicillin G into 6-APA and phenylacetic acid. A penicillin acylase was described several years ago in *P. chrysogenum* [65]. It was initially though that the 6-APA acyltransferase and penicillin acylase activities were associated in a single enzyme [67]. However Gatenbeck and Brunsberg [66] partially purified a soluble acyl-CoA:6-APA acyltransferase and observed that it was not able to hydrolyze penicillin G. Similar results were obtained by Pruess and Johnson [60].

The possibility that the numerous industrial mutants of *P. chrysogenum* have different related activities can not be excluded.

The isopenicillin N acyltransferase described initially in crude extracts [68] that is able to directly convert isopenicillin N into penicillin G, copurify with 6-APA acyltransferase [61]. Isopenicillin N was however used inefficiently as substrate as compared to 6-APA.

All available evidence indicates that two enzyme activities are involved in the conversion of isopenicillin N into penicillin G (Fig. 7). An initial step appears to involve the cleavage of isopenicillin N to 6-APA. The presence of an isopenicillin N amidolyase (6-APA forming) has been suggested by Abraham [1] and Queener and Neuss [10] and it is consistent with the accumulation of α-aminoadipic acid [72, 73] and 6-oxopiperidine-2-carboxylic acid [74] in penicillin-producing cultures but the existence of such an enzyme as free form is not unequivocally proved. It is very likely that the 6-APA acyltransferase and the isopenicillin N amidolyase are associated forming an enzyme complex [61]. According to this hypothesis the transacylation of isopenicillin N to penicillin G would require two enzyme activities that might be intimately associated. The loss of one cofactor or the denaturalization of the isopenicillin N amidolyase activity during purification may explain why the enzyme acylates 6-APA but it is inefficient in transacylating isopenicillin N.

4 Enzymes Involved Specifically in Cephalosporins and Cephamycin Biosynthesis

4.1 Isopenicillin Epimerase

In cephalosporin and cephamycin-producing microorganisms isopenicillin N is converted to penicillin N by isopenicillin N epimerase (IPNE) an enzyme that converts the L-α-aminoadipic acid side chain to the D configuration. This activity was discovered initially in extracts of *C. acremonium* [28, 75] and it appears to be lacking in extracts of *Penicillium*. Assays for epimerase activity require a procedure for the separation of penicillin N (product) and isopenicillin N (substrate). A procedure based on HPLC separation of tetraacetylglucose isothyocianate derivatives of these two penicillins has been described [76]. We and others have used an alternative approach based on a microbiological assay which makes use of the differential sensitivity of *E. coli* ESS 22-31 and *Micrococcus luteus* ATCC 9341 to the two isomers. *M. luteus* is sensitive to isopenicillin N but it is less sensitive to penicillin N [77]. *E. coli* ESS 22-31 is very sensitive to the penicillin N formed, but it is insensitive to isopenicillin N up to a concentration of 100 μg ml^{-1}.

The epimerase of *C. acremonium* is so labile that a complete characterization could not be carried out [78]. The presence of the epimerase in extracts of *C. acremonium* and its extreme lability has been confirmed by other research groups [79, 80]. It can be assayed only in freshly prepared cell-free extracts. Therefore, the protein has not been purified and very little progress has been made on the characterization of this enzymatic step of the cephalosporin biosynthetic pathway. The epimerase activity was present in the supernatant after centrifugation at $50,000 \times g$ and apparently does not require Fe^{2+}, ascorbic acid or ATP [80]; the inclusion in the buffer of pyridoxal-5-phosphate during washing of the cells and sonication stimulated product formation [78].

On the other hand *S. clavuligerus* [77] and *S. lactamdurans* [81] have more stable epimerases. The IPNE of *S. clavuligerus* has been purified 35-fold. It is not stimulated by the cofactors associated with ACV cyclization or ring expansion activities. In addition it is not affected by EDTA or incubation under anaerobic conditions, suggesting that it does not require Fe^{2+} or other metal ions at difference of the cyclase or expandase. Addition of pyridoxal phosphate to epimerase assays does not increase activity in cell free extracts of *S. clavuligerus* but purification of epimerase in absence of pyridoxal phosphate results in a rapid and irreversible loss of activity. Two pyridoxal phosphate antagonists cycloserine and isoniazid, at a final concentration of 10 mM have no effect on epimerase activity, but iproniazid (10 mM) slightly inhibits epimerase activity, and hydroxylamine at the same concentration strongly inhibits the enzyme. Gel filtration indicated that the epimerase of *S. clavuligerus* has a molecular weight of 60 kDa.

The epimerase of *S. lactamdurans* has been purified 88-fold and studied in more detail. Purification was achieved by a combination of ammonium sulfate precipitation, Sephadex G-75 gel filtration and DEAE-HPLC ion exchange chromatography, and the epimerase band has been identified as a protein with a molecular weight of about 60 kDa [Castro et al. unpublished]. The purified enzyme carried out a time and protein-dependent racemization reaction without addition of cofactors. The reaction requires only isopenicillin N and enzyme in Tris-HCl buffer. There was a narrow peak of optimal pH at 7.0 and optimal temperature was observed at 25 °C with a slightly lower activity at 20 °C and 30 °C. The epimerase of *S. lactamdurans* is inhibited 100% by Cu^{2+} and Hg^{2+}, 50% by Zn^{2+} and to a lower extent (16%) by Co^{2+} and Mn^{2+} (Table 5).

Table 5. Characteristics of isopenicillin N isomerases

Source	Stabilizers	Inhibitors	Km	Molecular weight kDa	Refs.
C. acremonium	Pyridoxal-5-phosphate	ND^a	ND	ND	[79, 80]
S. clavuligerus	Pyridoxal-5-phosphate	Iproniazid hydroxylamine	ND	60	[77]
S. lactamdurans	Pyridoxal-5-phosphate	Cu^{2+}, Hg^{2+} Zn^{2+}, Co^{2+} Mn^{2+}	0.27 mM	59	[81]

^{a} ND: not determined

The apparent Km for the substrate, isopenicillin N, was 0.27 mM. The epimerase activity of *S. lactamdurans* was reversible, also converting penicillin N into isopenicillin N [Castro et al. unpublished]. The enzymatic was protected by addition of pyridoxal-phosphate during purification as described for the enzyme of *S. clavuligerus*. However the enzyme does not appear to require this cofactor which probably plays a role in stabilizing the enzyme by tightly binding to the protein during purification.

4.2 Deacetoxycephalosporin C Synthase (Expandase)

Penicillin N is converted to deacetoxycephalosporin C (DAOC) by the deacetoxy-cephalosporin C synthetase, the so-called ring expanding enzyme or expandase [82]. This enzyme was studied initially in *C. acremonium* [83, 79] and later in *S. clavuligerus* [84] and *S. lactamdurans* [85]. In all these organisms the DAOC synthase carries out the oxidative conversion of the five membered intermediate penicillin N into the six-membered deacetoxycephalosporin C (Fig. 9). The enzyme of *C. acremonium* requires

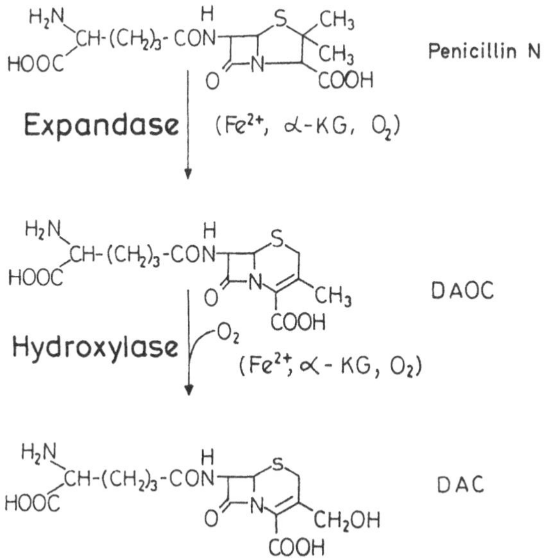

Fig. 9. Conversion of penicillin N into deacetoxycephalosporin C (DAOC) by the ring expanding enzyme (expandase or DAOC synthase) and hydroxylation of DAOC to give deacetylcephalosporin C (DAC) by DAOC-hydroxylase. The second reaction proceeds with incorporation of an oxygen atom from molecular oxygen

oxygen, α-ketoglutarate, Fe^{2+}, ascorbic acid and ATP [86, 70]. The DAOC synthase of *S. clavuligerus* does not seem to require ATP (or it does so to a very low extent) or ascorbic acid [84]. The enzyme of *S. lactamdurans* requires α-ketoglutarate, oxygen and Fe^{2+} but does not need ATP, ascorbic acid, Mg^{2+} or K^+ ions [85]. However due to the instability of the DAOC synthases of *C. acremonium* and *S. clavuligerus* and the inherent difficulty in purifying them [87, 88] some uncertainty concerning the required cofactors has remained. Using highly purified DAOC synthases from *C. acremonium* different research groups have provided additional evidence indicating that ATP may be stimulatory but it is not strictly required as a cofactor [89, 90]. The ATP stimulation may be an indirect functional property of the enzyme [89]. Therefore, the enzyme should be named synthase as already reported for the enzyme of *S. lactamdurans* [85] rather than synthetase (ATP-requiring).

The absolute requirement for α-ketoglutarate of the DAOC synthases of *S. clavuli-gerus* or *S. lactamdurans* that has also been shown unequivocally in the fungal enzymes [87, 89, 90] indicates that the expandases are intermolecular dioxygenases which require α-ketoglutarate as a substrate [91]. Other examples of α-ketoglutarate-dependent dioxygenases are thymidine-2-hydroxylase [92] and thymine-7-hydroxyl-ase [93] and also the DAOC hydroxylase (see later) of *C. acremonium*. α-Ketoglutarate is stoichiometrically decarboxylated to succinate [87, 94]. The availability of sufficient α-ketoglutarate may, therefore, be a limiting factor for ring-expansion. The require-ment for α-ketoglutarate of the DAOC synthase of *S. lactamdurans* is rather specific since neither glutamate, nor succinate, oxalacetate, α-ketobutyrate, α-ketoisovalerate, α-ketoadipate or α-ketocaproate were used as cofactors. α-Ketobutyrate completely inhibited the enzyme activity and β-ketoadipate produced a 56% reduction, probably as a result of a competitive inhibition. α-Ketodipate was an alternative cofactor for the fungal expandase with about 30% of the activity of α-ketoglutarate. This would suggest that the fungal enzyme has a broader cofactor specificity. However, the authors pointed out that there was a possibility that contaminant transaminases provide α-ketoglutarate (from glutamate) since the study was done with a partially-purified enzyme [70, 87].

All the DAOC synthases show a high specificity for the nature of the side chain in the penicillin substrate. Penicillin N, but not isopenicillin N, penicillin G or 6-amino-penicillanic acid served as substrates. The expandases of *C. acremonium* [70, 87] and *S. clavuligerus* [84] showed the same specificity. However, the DAOC synthase of *S. lactamdurans* was not inhibited by penicillin G as opposed to the fungal enzyme. Studies on side chain specificity of the expandase system of *C. acremonium* indicated that a six carbon *N*-acyl side chain, terminating in a carboxyl group, permits reasonable penam to cephem conversion [94 b].

The DAOC synthases of *C. acremonium* and *S. lactamdurans* showed a wide pH range of activity (pH 5 to 11) [89, 85]. Maximum stability was at pH 9.0. Advantage was taken of these facts to purify the enzyme using a pH gradient [85]. The ring expan-sion activity is not associated with membrane systems in any of studied organisms.

There is some disagreement in the molecular weight of the expandase and iso-electric point of the expandase of *C. acremonium* between the data of different authors. Kupka et al. [70, 87] and Scheidegger et al. [95] reported a molecular weight for this enzyme of 31 and 33 kDa, respectively and a pI of 4.6, while a molecular weight of 40–41 kDa and a pI of 6.3 have been described recently for the same enzyme [89, 90]. The origin of these variations is not known; it might be due to strain differences or to in vitro proteolytic degradation. The gene encoding the DAOC synthase of *C. acremonium* has been cloned and the open reading frame encodes a protein of 332 amino acids with a molecular weight of 36,462 Da (Fig. 10). Post-translational modification may be at least partially responsible for the anomalous migration of DAOC synthase during SDS polyacrylamide gel electrophoresis since the polypeptide migrates with an apparent molecular weight of 40 to 41 kDa [36]. The expandases of *S. clavuligerus* and *S. lactamdurans* have molecular weight values of 29,5 [88] and 27 kDa, respectively [85] which are lower than the molecular weight of the fungal enzyme. However, the fungal enzyme has probably arisen by a fusion of the DAOC synthase and DAOC hydroxylase genes of *Streptomyces* (see below). In all cases the expandase appears to be a monomer since the relative molecular mass of the natural (non-denatured) form, calculated from gel filtration is identical to that of the SDS denatured protein estimated by SDS-PAGE.

There are also some differences in the Km of the fungal and *S. lactamdurans* enzymes. The apparent Km of the *C. acremonium* enzyme is 29 µM for penicillin N [70,87,89] and 22 µM for α-ketoglutarate [89], while the *S. lactamdurans* enzyme showed an apparent Km of 52 µM for penicillin N, 3 µM for α-ketoglutarate and

```
                                                                    46
    T TTC GTT CTC ACT GGG ATC TTG TGA ATC CTT AAA TTC CTC TTG CAG
                                                                    94
AAC TTT CCT CCA CGC TAC TCC TCT CAA GTC ATC GCT CAA AAC CAC AGC
                                                                    142
ATC AAC ATG ACT TCC AAG GTC CCC GTC TTT CGT CTC GAC GAC CTC AAG
        met thr ser lys val pro val phe arg leu asp asp leu lys
                                                                    190
AGC GGC AAG GTC CTC ACC GAG CTC GCC GAG GCC GTC ACC ACC AAG GGT
ser gly lys val leu thr glu leu ala glu ala val thr thr lys gly
                                                                    238
ATC TTC TAC TTG ACC GAG AGC GGC CTG GTC GAC GAC GAC CAC ACC TCG
ile phe tyr leu thr glu ser gly leu val asp asp asp his thr ser
                                                                    286
GCG CGT GAG ACG TGC GTT GAC TTT TTC AAG AAC GGA AGC GAG GAG GAG
ala arg glu thr cys val asp phe phe lys asn gly ser glu glu glu
                                                                    334
AAG AGG GCC GTG ACG CTC GCC GAC CGT AAC GCC CGC CGC GGC TTC TCT
lys arg ala val thr leu ala asp arg asn ala arg arg gly phe ser
                                                                    382
GCC CTC GAG TGG GAG AGC ACC GCC GTC GTC ACC GAG ACG |GGC AAG TAC
ala leu glu trp glu ser thr ala val val thr glu thr |gly lys tyr
                                                                    430
TCG GAC TAC TCG ACG TGC TAC| TCC ATG GGC ATC GGC GGC AAC CTG TTC
ser asp tyr ser thr cys tyr| ser met gly ile gly gly asn leu phe
                                                                    478
CCG AAC CGG GGC TTC GAG GAC GTC TGG CAG GAC TAC TTC GAC CGC ATG
pro asn arg gly phe glu asp val trp gln asp tyr phe asp arg met
                                                                    526
TAC GGC GCA GCC AAG GAT GTC GCG CGC GCC GTT CTC AAC TCT GTG GGC
tyr gly ala ala lys asp val ala arg ala val leu asn ser val gly
                                                                    574
GCC CCG CTC GCC GGG GAG GAC ATT GAT GAC TTC GTC GAG TGC GAT CCC
ala pro leu ala gly glu asp ile asp asp phe val glu cys asp pro
                                                                    622
CTC CTC CGC CTA CGG TAC TTC CCG GAA GTG CCG GAG GAC CGC GTC GCC
leu leu arg leu arg tyr phe pro glu val pro glu asp arg val ala
                                                                    670
GAA GAG GAA CCC CTC CGC ATG GGA CCC CAC TAC GAC CTA TCG ACC ATC
glu glu glu pro leu arg met gly pro his tyr asp leu ser thr ile
                                                                    718
ACG CTC GTG CAC CAG ACA GCC TGC GCC AAC GGC TTC GTG AGC CTG CAG
thr leu val his gln thr ala cys ala asn gly phe val ser leu gln
                                                                    766
TGC GAG GTG GAC GGA GAA TTC GTC GAC CTC CCG ACG CTC CCC GGC GCC
cys glu val asp gly glu phe val asp leu pro thr leu pro gly ala
                                                                    814
ATG GTC GTC TTC TGC GGC GCG GTC GGC ACC CTG GCC ACG GGC GGC AAG
met val val phe cys gly ala val gly thr leu ala thr gly gly lys
                                                                    862
GTC AAG GCG CCC AAG CAC CGG GTC AAG TCT CCC GGG CGC GAC CAG CGC
val lys ala pro lys his arg val lys ser pro gly arg asp gln arg
                                                                    910
GTC GGC AGC AGC CGC ACG TCG AGC GTC TTC TTC CTG CGG CCG AAG CCC
val gly ser ser arg thr ser ser val phe phe leu arg pro lys pro
                                                                    958
GAC TTC AGC TTC AAC GTG CAG CAG TCG AGG GAG TGG GGT TTC AAC GTC
asp phe ser phe asn val gln gln ser arg glu trp gly phe asn val
                                                                    1006
CGC ATC CCG TCG GAG CGG ACG ACG TTC AGG GAG TGG CTT GGC GGC AAC
arg ile pro ser glu arg thr thr phe arg glu trp leu gly gly asn
                                                                    1054
TAT GTC AAC ATG CGG AGG GAT AAG CCG GCG GCA GCG GAG GCG GCT GTC
tyr val asn met arg arg asp lys pro ala ala ala glu ala ala val
                                                                    1102
CCC GCG GCT GCC CCT GTC TCT ACC GCA GCT CCT ATA GCC ACT TAG GGA
pro ala ala ala pro val ser thr ala ala pro ile ala thr
                                                                    1150
ACC CGC CGA TCG AGT AAT AAA TCT ACG GGA GTT TAA GAA GAA AAA TTG
                                                                    1198
CCC TAT AAA TTG CTA AAT TTT TAA AAC ACA AAG CAT GAG TGT CAA GAG
              1211
TTT CAA GTT TCA A
```

Fig. 10. Nucleotide and amino acid sequence of the DAOC-synthase/hydroxylase of *C. acremonium*. This gene encodes both enzymatic activities. The boxed region is very similar to a nucleotide sequence around the first cysteine amino acid that exists in the four cyclases (see the box in Fig. 4)

Table 6. Characteristics of deacetoxycephalosporin C synthases

Source	Activity	Requirements	Stimulators	Optimum pH	Optimum temp.	Km (μM)			Vmax(PenN) pmoles s^{-1} mg^{-1}	Molecular weight kDa	Inhibitor	Refs.
						PenN	αKG	DAOC				
C. acremonium	Expandase and hydroxylase	αKG Fe^{2+}	DTT ascorbate	7.3 to 7.8	26 to 34	29	22	—	13	41	N-Ethylmaleimide O-phenantroline DTNB, pHMB	89)
		O$_2$	ATP	7.3	36 to 38	—	22	20	2.8		Co^{2+}, Zn^{2+}, Mn^{2+}, CO$_3$NH$_4$ α-ketoadipate	90)
S. lactamdurans	Expandase	αKG Fe^{2+} O$_2$	DTT ascorbate	5 to 11	25 to 30	52	3	—		27	Mn^{2+}, Co^{2+}, Zn^{2+} α-ketobutyrate β-ketoadipate	85)
S. clavuligerus	Expandase	αKG Fe^{2+} K$^+$	ND^a	ND	ND	ND	ND	ND	ND	29,5	ND	88)
	Hydroxylase	Mg^{2+} ascorbate	ND	ND	ND	ND	ND	ND	ND	26,2	ND	138)

^{a} ND: not determined

71 μM for Fe^{2+} [85]. The Km for α-ketoglutarate of the DAOC synthase of *S. lactamdurans* is an order of magnitude lower (i.e. the enzyme shows more affinity) than the Km value of the *C. acremonium* enzyme (Table 6).

4.3 Deacetoxycephalosporin C Hydroxylase: Association with the Expandase

After the ring expansion of penicillin N to DAOC this compound is hydroxylated by an α-ketoglutarate-requiring dioxygenase to give deacetylcephalosporin C (DAC) [96]. This reaction is carried out by extracts of *C. acremonium* [97,98], *S. clavuligerus* [88,96] and *S. lactamdurans* [85]. This enzyme named DAOC hydroxylase or DAC synthase that catalyzes incorporation of oxygen from O_2 into DAOC (Fig. 9) [99], requires α-ketoglutarate, ascorbate, DTT and Fe^{2+}; i.e. DAOC hydroxylase and DAOC synthase are both intermolecular dioxygenases having similar cofactor requirements. The DAOC hydroxylase from *C. acremonium*, but not that from *S. clavuligerus*, is activated about 10-fold by preincubation with a mixture of Fe^{2+} and DTT. Fe^{2+} ions seem to play a key role in the activation [96].

The DAOC hydroxylase has a rather high substrate specificity. Substitution of a formyl, phenylacetyl, phenoxyacetyl, glutarimido, 2-(2-furyl)-2-methoxyiminoacetyl group or hydrogen for the D-α-aminoadipyl of DAOC give compounds which do not serve as substrates of the enzyme [10].

Scheidegger and co-workers [95] were unable to separate the two enzyme activities in extracts of an industrial strain of *C. acremonium* by several steps of purification and proposed, on this basis, that DAOC synthetase and DAOC hydroxylase are located on a single protein of a molecular weight of 33 kDa. The bifunctional role of this protein is supported by the well known result that *C. acremonium* accumulates only low concentrations of free deacetoxycephalosporin C. However, the bifunctional role could be due to two linked but different enzyme activities since mutants blocked in the conversion of DAOC to DAC have been isolated that accumulate increased levels of DAOC [100,101]. However, all these mutants are only partially blocked and still produce a considerable amount of DAC, suggesting that the expansion and oxygenation steps are catalyzed by a single enzyme. Recently, both DAOC synthase and hydroxylase activities have been found to remain physically associated and in a constant ratio of 7:1 when purified to near homogeneity from *C. acremonium*. The two activities can not be separated by ion exchange, dye ligand, gel filtration or hydrophobic chromatography [90]. The copurified expandase/hydroxylase appeared to be monomeric with a molecular weight of 41 +/−2 kDa and an isoelectric point of 6.3 [89,90]. The bifunctional nature of the enzyme is supported by the following observations: a) Chromatographic and electrophoretic inseparability, b) common requirement for α-ketoglutarate, Fe^{2+} and O_2, and stimulation by DTT, ascorbate and ATP, c) inhibition by metal chelators and sulphydryl reagents, including Zn^{2+}, d) similar pH and temperature stability patterns and e) very similar Km values for the three substrates (Table 5). There are however quantitative differences for the two enzymes in the optimal temperature, in sensitivity to Mn^{2+} and N-ethylmaleimide, in the stimulation by DTT and ATP and in the reversibility of DTNB inhibition.

A tight stoichiometric conversion (1:1) of penicillin N to DAOC + DAC by the expandase/hydroxylase of *C. acremonium* has been found. Moreover, DAOC is stoichiometrically converted to DAC (Fig. 11). These results suggest that the sequence of the two conversion is penicillin N → DAOC → DAC, and that there is no alternative pathway.

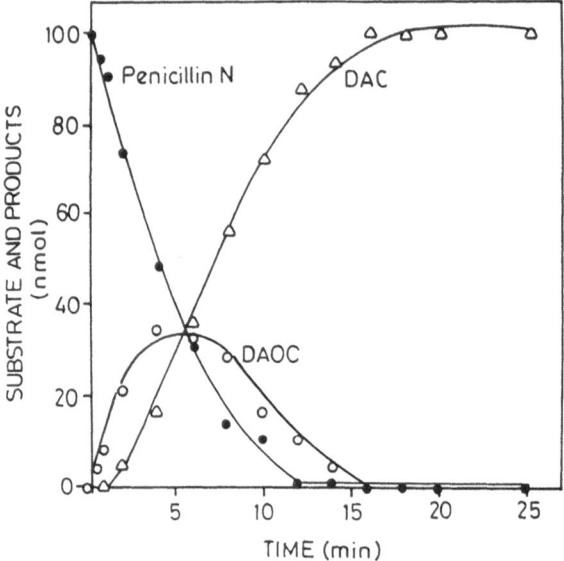

Fig. 11. Kinetics of the in vitro stoichiometric conversion of penicillin N into DAC by the DAOC synthase/hydroxylase. Note the transient accumulation of DAOC as an intermediate in the reaction (redrawn from Ref. [89])

In contrast to the association between the DAOC synthase and DAOC hydroxylase activities in *Cephalosporium*, both activities have been partially purified from extracts of *S. clavuligerus* and separated by ion exchange chromatography. A molecular weight of 26,2 kDa was estimated for the DAOC hydroxylase of *S. clavuligerus* which is slightly smaller than the estimated value of 29,5 kDa for the DAOC synthase of the same organisms [88]. The highly purified expandase of *S. lactamdurans* showed no DAOC hydroxylase activity, which is however present in cephamycin synthesizing cells, suggesting that the two activities are separated during purification [85].

In conclusion, in cephamycin producing *Streptomyces*, there appear to be two separated enzymes carrying out the expansion and hydroxylation activities whereas in *Cephalosporium* all the evidence supports the existence of a bifunctional enzyme, or two very similar enzymes, which cannot be separated by chromatographic procedures.

The bifunctional nature of the DAOC synthase hydroxylase polypeptide and the cofactor/substrate requirements shared by DAOC synthase and DAOC hydroxylase, raise the question of whether a single site or two active sites are responsible for the two activities in *A. chrysogenum*.

4.3.1 A Fused Gene for DAOC Synthase/DAOC Hydroxylase in *Cephalosporium*

Cloning of the structural gene for the expandase and hydroxylase and the expression into *E. coli* of the cloned gene has provided unambiguous proof that the expandase and hydroxylase are encoded by a single polypeptide [36]. The *C. acremonium* DAOC synthase/hydroxylase fused gene may have evolved by fusion of two contiguous separate genes encoding each of the enzymes in the genome of *Streptomyces*. Comparison of the protein coding regions of the two genes of *Streptomyces* to the protein coding region of the bifunctional *C. acremonium* gene will lead to the identification of the two domains within the *C. acremonium* gene. The possibility that the separable DAOC synthase and DAOC hydroxylase activities in *Streptomyces* arise from post-translational modification of a single polypeptide seems unlikely although it cannot be ruled out at the present time.

4.3.2 Molecular Mechanisms of Ring Expansion and DAOC Hydroxylation

The molecular mechanism of the ring expansion and hydroxylation reactions is not clear. Since there is no net introduction of oxygen during the ring expansion, a transient oxygen addition appears to be followed by a deoxygenation step. In analogy with the chemical conversion of penicillins into cephalosporins, the β-sulphoxide derivative has been proposed as a potential intermediate in the biosynthetic transformation of penicillin N into DAOC (Fig. 11) [10, 102]. Alternatively, the β-methylene-hydroxy-penam has also been proposed as a feasible intermediate for the conversion [5, 102]. Both postulates are in stereochemical agreement with the observation that the 3-β-methyl group of penicillin N is incorporated into the dihydrothiazine moiety of DAOC (Fig. 12) [103, 104]. Neither compound was, however, a substrate for the enzyme, suggesting that these are not free intermediates in the reaction [90]. This result does not rule out the possibility of the formation of one of them as an enzyme

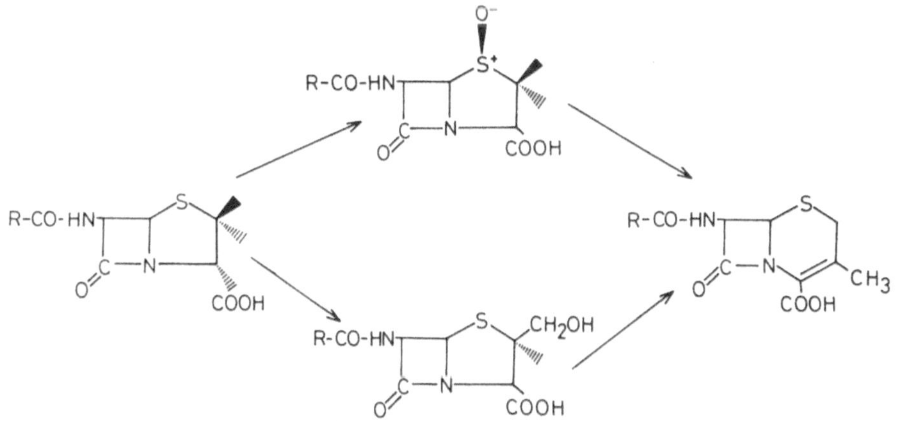

Fig. 12. Molecular mechanisms for the conversion of penicillin N into DAOC. Two alternative compounds, the β-sulfoxide derivative of penicillin N (upper) and a β-methylenehydroxypenam (lower), have been proposed as intermediates

bound intermediate. A synthetic analogue in which the 3-β-methyl group and the 2-hydrogen atom of penicillin N were replaced by a cyclo-propene ring was not a substrate but was a reversible inhibitor of the enzyme [90].

Functional sulphydryl groups and Fe^{2+} are involved in the catalysis by the expandase/hydroxylase. The protein appears to have at least one catalytically important thiol group. Both, the expandase and hydroxylase activities are inhibited by p-hydroxymercuribenzoate, N-ethylmaleimide, 5,5'-dithio-bis-2-nitrobenzoic acid and by Zn^{2+} which is known to bind sulphydryl groups [105].

DTT reactivates deacetoxycephalosporin C synthase after inactivation [106]. DTT and ascorbate contribute to the stability and optimal activity of the enzyme, presumably by keeping iron in the reduced form and by protecting essential sulphydryl groups of the enzyme. The activities of both expandase and hydroxylase of C. acremonium are greatest with Fe^{2+} and Fe^{3+} in the presence of DTT and ascorbate, suggesting that one function of DTT and ascorbate is iron reduction [89]. The same observation has been made using the purified expandase of S. lactamdurans [85].

4.3.3 Control of the Cephalosporin and Cephamycin Biosynthesis at the Expandase Level

Under normal process conditions, the ring expansion reaction appears to be limiting for cephalosporin or cephamycin production. The intermediates penicillin N, DAOC and DAC are released into the culture medium in varying amounts depending on cultivation conditions and mutants used [95]. A decreased cephalosporin C production and accumulation of penicillin N is observed after addition of glucose or ammonium to cultures of C. acremonium [107]. In effect, the expandase of C. acremonium and S. lactamdurans are repressed by both carbon and nitrogen sources [18, 108, 110]. Jensen and co-workers [111] have reported that oxygen transfer is limiting for expandase activity in S. clavuligerus thereby limiting cephamycin biosynthesis. All this evidence supports the conclusion that ring expansion represents one of the rate-limiting steps of cephalosporin synthesis [112].

4.4 Deacetylcephalosporin C Acetyltransferase

Acetylation of DAC to cephalosporin C by the enzyme acetyl-CoA:deacetylcephalosporin C acetyltransferase is the terminal reaction in the cephalosporin producing fungi (Fig. 13) [113]. A similar enzyme probably exists in Streptomyces lipmanii which produces 7-α-methoxycephalosporin C in addition to DAC [114, 115]. Cephalosporin C negative mutants which accumulate DAC have been isolated [101, 116]. Extracts of the wild type C. acremonium, but not those of cephalosporin-negative mutants, are able to convert DAC into CPC in the presence of acetyl-CoA and Mg^{2+} [117]. Other divalent cations, Zn^{2+}, Cd^{2+}, Mn^{2+}, Ni^{2+}, Pb^{2+} and Cu^{2+} also support enzyme activity although it is lower than with Mg^{2+} at the same concentration (1 mM). The optimal pH of the enzyme was 7.0 to 7.5 [113].

Permeabilized cells of C. acremonium obtained by treatment with ether are able to take DAC and convert it into CPC in the presence of acetyl-CoA [118]. The acetyl CoA:DAC O-acetyltransferase has been purified 10-fold by ammonium sulfate fractionation and gel filtration [101]. The enzyme has a high substrate affinity for the

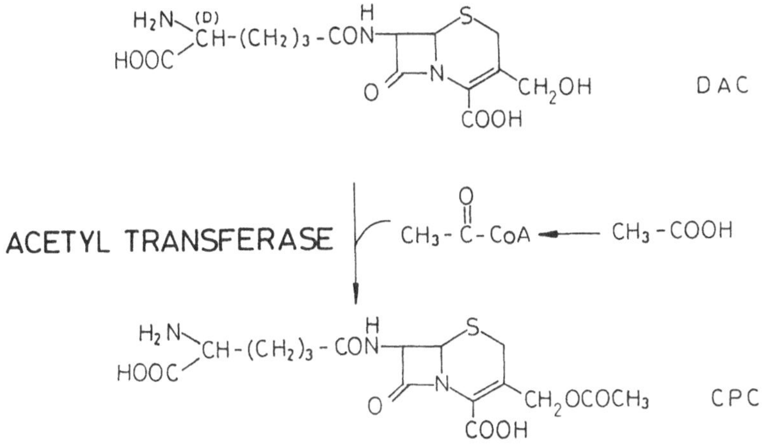

Fig. 13. Conversion of DAC into cephalosporin C by the enzyme Acetyl-CoA:deacetylcephalosporin C acetyltransferase

hydroxymethyl group at carbon 3 of DAC. The hydroxymethyl groups of DL-serine or DL-homoserine are not acetylated by this enzyme.

One of the mutants blocked in the conversion of DAC to CPC accumulated a new compound in which the hydroxymethyl is converted into an aldehyde [113] which suggests that when the acetylation can not proceed the DAOC hydroxylase is able to further oxydize DAC to the corresponding aldehyde.

The CPC formed may be degraded back to DAC by an extracellular acetylhydrolase in some strains of *C. acremonium* [117, 119]. Fortunately, this enzyme is repressed by glucose and many other carbohydrates and usually does not represent a major problem in CPC bioprocesses [120].

5 Late Reactions in Cephamycin Biosynthesis

Cephalosporin C is the end-product of the biosynthetic pathway in *C. acremonium*. However, further reactions are involved in the formation of the C-7 methoxy group and in the attachment of the carbamoyl group at C-3 during cephamycin C biosynthesis in the actinomycetes.

5.1 Conversion of Cephalosporin C into 7-α-Methoxycephalosporin C. Cephalosporin C Hydroxylase and 7-α-Hydroxycephalosporin C Methyltransferase

The methoxy group at C-7 of cephamycins derives from molecular oxygen and methionine and is introduced by an enzyme system containing a dioxygenase and a methyltransferase (Fig. 14) [121, 122]. Crude extracts of *S. clavuligerus* in the presence

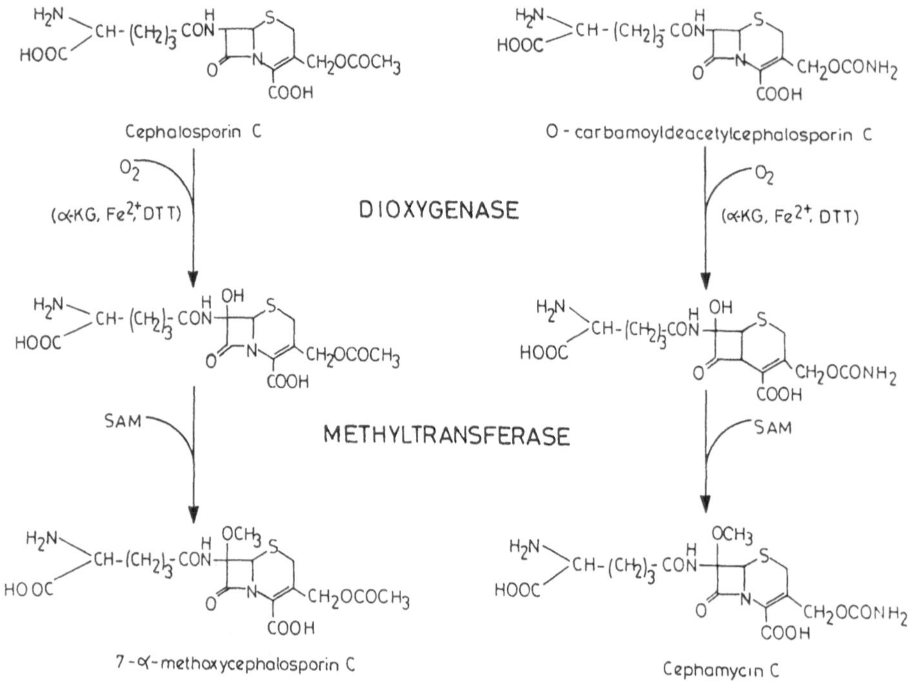

Fig. 14. Conversion of cephalosporin C into 7-α-methoxycephalosporin C (*left*) by cephalosporin C hydroxylase and 7-α-hydroxycephalosporin C methyltransferase. The same enzymes are able to convert *O*-carbamoyl deacetylcephalosporin C (a cephamycin intermediate) into cephamycin C

of S-adenosylmethionine, 2-α-ketoglutarate, Fe^{2+} and a reducing agent convert cephalosporin C or *O*-carbamoyldeacetylcephalosporin C into 7-α-methoxy derivatives [123]. No synthesis of a 7-α-methoxyderivatives of deacetylcephalosporin C is observed, and the 7-α-methoxyderivatives of deacetoxycephalosporin C was produced only in small amounts. Based on these results O'Sullivan and Abraham [123] concluded that the 7-α-methoxy group is introduced after the cephalosporin C molecule has been formed. Evidence was later presented that, in *S. clavuligerus*, the methoxylation reaction proceeds by a two step reaction involving the formation of 7-α-hydroxy-cephalosporin C [124] with subsequent methylation to yield 7-α-methoxycephalo-sporin C using *S*-adenosylmethionine as a methyl donor [124]. The two reactions could be separated by excluding *S*-adenosylmethionine from the reaction mixture.

5.2 Carbamoylation of Deacetylcephalosporin C by *O*-Carbamoyltransferase

One of the final products accumulated in processes of *S. clavuligerus* is *O*-carbamoyl-deacetylcephalosporin C [115, 125] in which the acetyl group of cephalosporin C has been replaced by a carbamoyl residue. The reaction is carried out by a *O*-carbamoyl-transferase present in cells of *Streptomyces clavuligerus* which are actively synthesizing cephamycin C [98]. The enzyme transfers a carbamoyl group from carbamoylphosphate

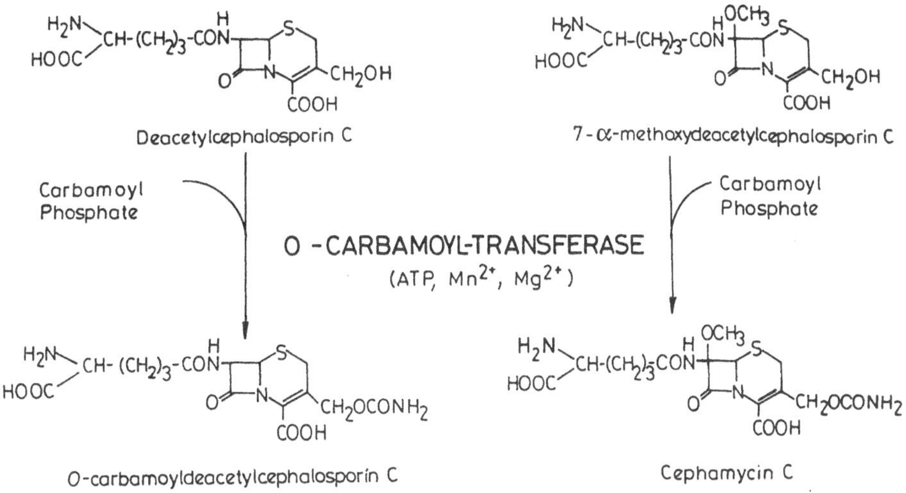

Fig. 15. Conversion of deacetylcephalosporin C into *O*-carbamoyldeacetylcephalosporin C (*left*) by *O*-carbamoyl transferase. The same enzyme converts 7-α-methoxydeacetylcephalosporin C into cephamycin C (*right*)

to a 3-hydroxymethylceph-3-em-4-carboxylic acid nucleus to form a 3-carbamoyl-oxymethylcephem [126] (Fig. 15). Similar enzymes are involved in the transfer of carbamoyl groups in the biosynthesis of mitomycins and polyoxins.

The enzyme of *S. clavuligerus* has been purified 40-fold by batch absorption onto DEAE-cellulose and hydroxyapatite chromatography. The purified *O*-carbamoyltrans-ferase is most active at pH 6.8. The enzyme is stimulated by Mg^{2+} and Mn^{2+}, and also by ATP but it is not known whether this nucleotide acts as an effector or as a substrate. Some activity is observed with dATP but other ATP analogues inhibit the action of ATP itself. The enzyme activity is stabilized by phosphate anions but it is inhibited by pyrophosphate anions.

The *O*-carbamoyltransferase synthesizes a wide variety of 3-carbamoyloxymethyl-cephems since it tolerates structural alterations around the 7-amino group of cephalo-sporins in contrast to the highly specific DAOC hydroxylase which oxidizes the 3-methyl group of DAOC to a 3-hydroxymethyl group to form DAC. Even 7-amino deacetylcephalosporanic acid is a substrate of the carbamoyltransferase even though those cephalosporins that carry 7-amino side chains with no net charges are the best substrates [126].

Crude homogenates of *S. clavuligerus* can synthesize cephamycin C from 7β-aminoadipoamido-3-hydroxymethyl-7α-methoxyceph-3-em-4-carboxylic acid and ci-trulline. However it is not clear whether a direct transfer of the carbamoyl-group occurs since the direct esterification with carbamoyl-phosphate is more efficient [126]. It is likely that in the presence of phosphate, citruline is converted to carbamoylphosphate (the direct precursor) and ornithine by the action of ornithine transcarbamylase.

This hypothesis is consistent with the observation that ornithine (one of the products of the ornithine transcarbamylase) drastically inhibits cephamycin [127] presumably by inhibiting the splitting activity that generates carbamoylphosphate.

5.3 Future Outlook

Basic studies on biosynthesis of β-lactam antibiotic have progressed slowly for many years [128]. However, the recent application of molecular biology techniques to *P. chysogenum, C. acremonium, A. nidulans, S. clavuligerus* and *S. lactamdurans* has provided a new spirit to the study of β-lactam antibiotics. Cloning of the isopenicillin N synthase and deacetoxycephalosporin C synthase has been achieved, and others are in progress [59]. Availability of cloned genes opens the doors for in vitro mutagenesis and in vivo expression studies. Efficient transformation vectors for these micro-organism have been developed [129]. Amplification of the available genes will help to elucidate bottlenecks in the β-lactam antibiotics biosynthetic pathways [59]. Knowledge of the promoters and efficient gene expression will have a positive influence in the industrial production of penicillins and cephalosporins.

6 Acknowledgements

This work supported by grants of the CICYT, Madrid, and Gist-Brocades, Delft, The Netherlands. We thank M. I. Corrales for typing the manuscript.

7 References

1. Abraham EP (1986) Enzymes involved in penicillin and cephalosporin formation. In: Kleinkauf H, Dohren H, Dornauer H, Nesemann G (eds) Regulation of secondary metabolite formation. VCH Verlag, Weinheim, p 115
2. Enriquez L, Pisano MA (1979) Antimicrob. Ag. Chemother. *16*: 392
3. López-Nieto MJ, Ramos FR, Luengo JM, Martin JF (1985) Appl. Microbiol. Biotechnol. *22*: 343
4. Adriaens P, Meeschaert B, Wuyts W, Vanderhaegue H, Eyssen H (1975) Antimicrob. Ag. Chemother. *8*: 638
5. Abraham EP (1977) J. Antibiot. *30*: S1
6. Demain AL (1983) Biosynthesis of β-lactam antibiotics. In: Demaiu AL, Solomon NA (eds) Antibiotics containing the β-lactam structure. Springer, Berlin Heidelberg New York, vol 1 p 189
7. Fawcett PA, Abraham EP (1975) Methods in Enzymol. *42*: 471
8. Katz E, Weissbach H (1963) J. Biol. Chem. *238*: 666
9. Ramos FR, López-Nieto MJ, Martín JF (1985) Antimicrob. Ag. Chemother. *27*: 380
10. Queener S, Neuss N (1982) The biosynthesis of β-lactam antibiotics. In: Morin RB, Morgan M (eds) The chemistry and biology of β-lactam antibiotics. Academic, London New York, p 1
11. Delderfield JS, Mtetwa E, Thomas R, Tyobeka TE (1981) J. Chem. Soc. Chem. Comm. 650
12. Fawcett PA, Usher JJ, Huddleston JA, Bleany RC, Nisbet JJ, Abraham EP (1976) Biochem. J. *157*: 651
13. Lara F, Mateos RC, Vázquez G, Sánchez S (1982) Biochem. Biophys. Res. Commun. *105*: 172
14. Adlington M, Baldwin JE, López-Nieto MJ, Murphy JF, Patel NA (1983) Biochem. J. *213*: 573
15. López-Nieto MJ (1984) Ph. D. Thesis, Universidad de Salamanca
16. Banko G, Wolfe S, Demain AL (1986) Biochem. Biophys. Res. Commun. *137*: 528
17. Banko G, Demain AL, Wolfe S (1987) J. Am. Chem. Soc. *109*: 2858
18. Cortés J, Liras P, Castro J, Martín JF (1986) J. Gen. Microb. *132*: 1805
19. Jensen SE, Westlake DWS, Wolfe S (1988) FEMS Microbiol. Lett. *49*: 213

20. Loder PB, Abraham EP (1971) Biochem. J. *123*: 471
21. Pang C-P, Chakravarti B, Adlington RM, Ting H-H, White RL, Jayatilake GS, Baldwin JE, Abraham EP (1984) Biochem. J. *222*: 789
22. Hollander IJ, Shen Y-Q, Heim J, Demain AL (1984) Science *224*: 610
23. Jensen SE, Leskiw BK, Vining LC, Aharonowitz Y, Westlake DWS, Wolfe S (1986) Can. J. Microbiol. *32*: 953
24. Castro JM, Liras P, Laiz L, Cortés J, Martín JF (1988) J. Gen. Microb. *134*: 133
25. Martín JF, Díez B, Alvarez E, Barredo JL, Cantoral MJ (1987) Development of a transformation system in *Penicillium chrysogenum*: Cloning of genes involved in penicillin biosynthesis. In: Alacevic M, Hranueli D, Toman Z (eds) Genetics of industrial microorganisms, Pliva, Zagreb, p 297
26. Jensen SE, Westlake DWS, Wolfe S (1982) J. Antibiot. *35*: 483
27. Revilla G, López-Nieto JM, Luengo JM, Martín JF (1984) J. Antibiot. *37*: 781
28. Konomi T, Herchen D, Baldwin JE, Yoshida M, Hunt NA, Demain AL (1979) Biochem. J. *184*: 427
29. O'Sullivan J, Bleaney RC, Huddleston JA, Abraham EP (1979) Biochem. J. *184*: 421
30. Baldwin JE, Johnson BL, Usher JJ, Abraham EP, Huddleston JA, White RL (1980) J. Chem. Soc. Chem. Commun. 1271
31. Bahadur GA, Baldwin JE, Usher JJ, Abraham EP, Jayatilake GS, White RL (1981) J. Am. Chem. Soc. *103*: 7650
32. Baldwin JE, Adlington RM, Ting HH, Arigone C, Graf P, Martinoni B (1985) Tetrahedron *41*: 3339
33. Baldwin JE, Wan TS (1981) Tetrahedron *37*: 1589
34. Easton CJ (1983) J. Chem. Soc. Chem. Commun. 1349
35. Baldwin JE, Gagnon J, Ting HH (1985) FEBS Lett. *188*: 253
36. Samson SM, Chapman JL, Belagaje R, Queener SW, Ingolia TD (1987) Proc. Natl. Acad. Sci. USA *84*: 1
37. Martín JF (1979) Biosynthesis of metabolic products with antimicrobial activities: β-lactam antibiotics. In: Schaal KP, Pulverer G (eds) Actinomycetes, Fisher, Stuttgart, p 417
38. Samson SM, Belagaje R, Blankenship DT, Chapman JL, Perry D, Skatrud PL, Vanfrank RM, Abraham EP, Baldwin JE, Queener SW, Ingolia TD (1985) Nature *318*: 191
39. Chapman JL, Skatrud PL, Ingolia TD, Samson SM, Kaster KR, Queener SW (1987) Dev. Ind. Microbiol. *27*: 165
40. Baldwin JE, Killin SJ, Pratt AJ, Sutherland JD, Turner NJ, Crabbe JC, Abraham EP, Willis AC (1987) J. Antibiot. *40*: 652
40a. Barredo JL, Cantoral JM, Alvarez E, Diez B, Martín JF (1989) Mol. Gen. Genet. *216*: 91
41. Carr LG, Skatrud PL, Scheetz ME, Queener SW, Ingolia TD (1986) Gene *48*: 257
42. Ramón D, Carramolino L, Patiño C, Sánchez F, Peñalva MA (1987) Gene *57*: 171
42a. Leskiw BK, Aharonowitz Y, Mevarech M, Wolfe S, Vining LC, Westlake DWS, Jensen SE (1988) Gene *62*: 187
43. Kim K, Rhee SG, Stadthan ER (1985) J. Biol. Chem. *260*: 15394
44. Wolfe S, Demain AL, Jensen SE, Westlake DWS (1984) Science *226*: 1386
45. Demain AL, Kupka J, Shen YQ, Wolfe S (1982) Microbiological synthesis of β-lactam antibiotics. In: Umezawa H (ed) Trends in antibiotics research: Genetics, biosynthesis, action and new substances, Jap. Ant. Res. Ass., Tokyo, p 233
46. Baldwin JE, Abraham EP, Adlington RM, Chakravarti B, Derome AE, Murphy JA, Field LD, Green NB, Ting HH, Usher JJ (1983) J. Chem. Soc. Chem. Commun. 1317
47. Wolfe S, Hollander IJ, Demain AL (1984) Biotechnology *2*: 635
48. Bowers RJ, Jensen SE, Lyubechansky L, Westlake DWS, Wolfe S (1984) Biochem. Biophys. Res. Commun. *120*: 607
49. Jensen SE, Westlake DWS, Bowers RJ, Ingold CF, Jouany M, Lyubechansky L, Wolfe S (1984) Can. J. Chem. *62*: 2712
50. Baldwin JE, Adlington RM, Basak A, Flitsch SL, Pettursson S, Turner NJ, Ting HH (1986) J. Chem. Soc. Chem. Commun. 975
51. Baldwin JE, Abraham EP, Adlington RM, Bahadur GA, Chakravarti B, Domayne-Hayman BP, Field LD, Flitson SL, Jayatilake GS, Spakovskys A, Ting HH, Turner NJ, White RL, Usher JJ (1984) J. Chem. Soc. Chem. Commun. 1225

52. Luengo JM, Alemany MT, Salto F, Ramos F, López-Nieto MJ, Martín JF (1986) Biotechnology *4*: 44
53. Castro JM, Liras P, Cortés J, Martín JF (1986) FEMS Microbiol. Lett. *34*: 349
54. Jensen SE, Westlake DWS, Bowers RJ, Lyubechansky L, Wolfe S (1986) J. Antibiot. *39*: 822
55. Luengo JM, López-Nieto MJ, Salto F (1986) J. Antibiot. *39*: 1144
56. Baldwin JE, Abraham EP, Burge GL, Ting HH (1985) J. Chem. Soc. *24*: 1808
57. Jensen SE, Westlake DWS, Wolfe S (1984) Appl. Microbiol. Biotechnol. *20*: 155
58. Skatrud PL, Queener SW, Carr LG, Fisher DL (1987) Curr. Genet. *12*: 337
59. Martín JF (1987) Trends Biotechnol. *5*: 306
60. Pruess DL, Johnson MJ (1967) J. Bacteriol. *94*: 1502
61. Alvarez E, Cantoral JM, Barredo JL, Díez B, Martín JF (1987) Antimicrob. Ag. Chemother. *31*: 1675
62. Batchelor FR, Doyle FP, Nayler JHC, Rolinson GN (1959) Nature *183*: 257
63. Cole M (1966) Appl. Microbiol. *14*: 98
64. Kitano K, Kintaka K, Kamamoto K, Nara KS, Nakao Y (1975) J. Ferment. Technol. *53*: 339
65. Brunner R, Roehr M, Zinner M (1968) Hoppe-Seyler's Z. Physiol. Chem. *349*: 95
66. Gatenbeck S, Brunsberg U (1968) Acta. Chem. Scand. *22*: 1059
67. Spencer B, Maung C (1970) Biochem. J. *118*: 29
68. Fawcett PA, Usher JJ, Abraham EP (1975) Biochem. J. *151*: 741
69. Ramos FR, López-Nieto MJ, Martín JF (1986) FEMS Microbiol. Lett. *35*: 123
70. Kupka J, Shen Y-Q, Wolfe S, Demain AL (1983) Can. J. Microbiol. *29*: 488
71. Martín JF, Demain AL (1978) In: Smith JE, Berry DR (eds) The Filamentous Fungi, Edward Arnold, London, vol 3, p 425
72. Friedrich CG, Demain AL (1978) Arch. Microbiol. *119*: 43
73. Revilla G, Ramos FR, López-Nieto MJ, Alvarez E, Martín JF (1986) J. Bacteriol. *168*: 947
74. Brundidge SP, Gaeta FCA, Hook DJ, Sapino C, Elander RP, Morin RB (1980) J. Antibiot. *35*: 1348
75. Sawada Y, Konomi T, Solomon N, Demain AL (1980) FEMS Microbiol. Lett. *9*: 281
76. Neuss N, Berry DM, Kupka J, Demain AL, Queener SW, Duckworth DC, Huckstep LL (1982) J. Antibiot. *35*: 580
77. Jensen SE, Westlake DWS, Wolfe S (1983) Can. J. Microbiol. *29*: 1526
78. Lubbe C, Wolfe S, Demain AL (1986) Appl. Microb. Biotechnol. *23*: 367
79. Baldwin JE, Deeping JW, Singh PD, Vallejo CA (1981) Biochem. J. *194*: 649
80. Jayatilake GS, Huddleston JA, Abraham EP (1981) Biochem. J. *194*: 645
81. Castro JM (1985) Ph. D. Thesis, Universidad de León
82. Koshaka M, Demain AL (1976) Biochim. Biophys. Res. Commun. *70*: 465
83. Yoshida M, Konomi T, Kohsaka M, Baldwin JE, Herchen S, Singh P, Hunt NA, Demain AL (1978) Proc. Natl. Acad. Sci. U.S.A. *75*: 6253
84. Jensen J, Westlake DWS, Bowers RJ, Wolfe S (1982) J. Antibiot. *34*: 1351
85. Cortés J, Martín JF, Castro JM, Laiz L, Liras P (1987) J. Gen. Microbiol. *133*: 3165
86. Hook DJ, Chang JT, Elander RP, Morin RB (1979) Biochem. Biophys. Res. Commun. *87*: 258
87. Kupka J, Shen YQ, Wolfe S, Demain AL (1983) FEMS Microbiol. Lett. *16*: 1
88. Jensen SE, Westlake DWS, Wolfe S (1985) J. Antibiot. *38*: 263
89. Dotzlaf JE, Yeh WK (1987) J. Bacteriol. *169*: 1611
90. Baldwin JE, Adlington RM, Coates JB, Crabbe JC, Crouch NP, Keeping JW, Knight GC, Schofield CJ, Ting HH, Vallejo CA, Thorniley M, Abraham EP (1987) Biochem. J. *245*: 831
91. Abbot MT, Underfriend S (1974) α-Ketoglutarate-coupled dioxygenases. In: Hayashi O (ed) Molecular mechanisms of oxygen activation, Academic, London, p 167
92. Bankel L, Lindstedt G, Lindstedt S (1972) J. Biol. Chem. *247*: 6128
93. Liu CK, Hsu CA, Abbot MT (1973) Arch. Biochem. Biophys. *159*: 180
94. Baldwin JE, Crabbe MJC (1987) FEBS Letters *214*: 2357
94b. Baldwin JE, Adlington RM, Crabbe J, Knight G, Nomoto T, Schofield C, Ting H (1987) Tetrahedron *43*: 3009
95. Scheidegger A, Küenzi MT, Nüesch J (1984) J. Antibiot. *37*: 522
96. Turner MK, Farthing JE, Brewer SJ (1978) Biochem. J. *173*: 839
97. Fujisawa Y, Kikuchi M, Kanzaki T (1977) J. Antibiot. *30*: 775

98. Brewer SJ, Boyle TT, Turner MK (1977) Biochem. Soc. Trans. *5*: 1026
99. Stevens CM, Abraham EP, Huang F-C, Sih CJ (1975) Fed. Proc. *34*: 625
100. Fujisawa Y, Kitano K, Kanzaki T (1975) Agr. Biol. Chem. *39*: 2049
101. Liersch M, Nüesch J, Treichler HJ (1976) Final steps in the biosynthesis of cephalosporin C. In: MacDonald KD (ed) International symposium on genetics of industrial microorganisms, Academic, London, p 179
102. Baldwin JE, Herchen SR, Clardy JC, Hirotsu K, Chou TS (1978) J. Org. Chem. *43*: 1342
103. Kluender H, Bradley CH, Sih CJ, Fawcett P, Abraham EP (1973) J. Am. Chem. Soc. *95*: 6149
104. Neuss N, Nash CH, Baldwin JE, Lemke PA, Grutzner JB (1973) C. J. Am. Chem. Soc. *95*: 3797
105. Mukerji SK, Pimstone NR (1986) Arch. Biochem. Biophys. *244*: 619
106. Lubbe C, Wolfe S, Demain AL (1985) Enzyme and Microb. Technol. *7*: 353
107. Zanca DM, Martín JF (1983) J. Antibiot. *36*: 700
108. Heim J, Shen JQ, Wolfe S, Demain AL (1984) Appl. Microbiol. Biotechnol. *19*: 232
109. Shen Y-Q, Heim J, Solomon NA, Wolfe S, Demain AL (1984) J. Antibiot. *37*: 503
110. Castro JM, Liras P, Cortés J, Martín JF (1985) Appl. Microbiol. Biotechnol. *22*: 32
111. Rollings MJ, Carmichael RD, Jensen SE, Westlake DWS (1987) Abs. SIM Meeting P-69, p 80
112. Martín JF, López-Nieto JM, Castro JM, Cortés J, Romero J, Ramos FR, Cantoral JM, Alvarez E, Domínguez MG, Barredo JL, Liras P (1986) Enzymes involved in β-lactam biosynthesis controlled by carbon and nitrogen regulation. In: Kleinkauf H, Döhren H, Dornauer H, Nesseman G (eds) Regulation of secondary metabolite formation, VCH, Weinheim, p 41
113. Fujisawa Y, Kanzaki T (1975) Agric. Biol. Chem. *39*: 2043
114. Higgens CE, Hamil, RL, Sands TH, Hoehn MM, Davis NE, Nagarajan R, Boeck LD (1974) J. Antibiot. *27*: 298
115. Nagarajan R, Boeck LD, Gorman M, Hamill RL, Higgens CE, Hoehn MM, Stark WM, Whitney JG (1971) J. Am. Chem. Soc. *93*: 2308
116. Fujisawa Y, Shirafuji H, Hida M, Nara K, Yoneda M, Kanzaki T (1975) Agr. Biol. Chem. *39*: 1295
117. Fujisawa Y, Shirafuji H, Hida M, Nara KL, Yoneda M, Kanzaki T (1973) Nature *246*: 154
118. Felix HR, Nüesch J, Wehrli W (1980) FEMS Microbiol. Lett. *8*: 55
119. Hinnen A, Nüesch J (1976) Antimicrob. Ag. Chemother. *9*: 824
120. Huber FM, Baltz RH, Caltrider PG (1968) Appl. Microbiol. *16*: 1011
121. O'Sullivan J, Applin RT, Stevens CM, Abraham EP (1979) Biochem. J. *179*: 47
122. Whitney JG, Brannon DR, Mabe JA, Wicker KJ (1972) Antimicrob. Ag. Chemother *1*: 247
123. O'Sullivan J, Abraham EP (1980) Biochem. J. *186*: 613
124. Hood JD, Elson A, Gilpin ML, Brown AG (1983) J. Chem. Soc. Chem. Commun. 1187–1188
125. Higgens CE, Kastner RE (1971) Int. J. Syst. Bacteriol. *21*: 326
126. Brewer SJ, Taylor PM, Turner MK (1980) Biochem. J. *185*: 555
127. Romero J, Liras P, Martín JF (1986) Appl. Env. Microbiol. *50*: 50
128. Martín JF, Liras P (1985) Trends Biotechnol. *3*: 39
129. Cantoral JM, Díez B, Barredo JL, Alvarez E, Martín JF (1987) Biotechnology *5*: 494
130. Díez B, Alvarez E, Cantoral JM, Barredo JL, Martín JF (1987) Curr. Genet. *12*: 277
131. García Domínguez M, Martín JF, Mahro B, Demain AL, Liras P (1987) Appl. Env. Microbiol. *53*: 1376
132. Baldwin JE, Abraham EP, Burge GL, Ting HH (1985) J. Chem. Soc. Chem. Commun. 1808
133. Baldwin JE, Adlington RM, Crabbe MJ, Knight GC, Nomoto T, Schofield CJ (1987) J. Chem. Soc. Chem. Commun. 806
134. Baldwin JE, Pratt AJ, Moloney MG (1987) Tetrahedron *43*: 2565
135. Shen YQ, Wolfe S, Demain AL (1984) J. Antibiot. *37*: 1044
136. Baldwin JE, Adlington RM, Domayne-Hayman BR, Ting HH, Turner NJ (1986) J. Chem. Soc. Chem. Commun. 110
137. Baldwin JE, Abraham EP, Lovel CG, Ting HH (1984) J. Chem. Soc. Chem. Commun. 902
138. Jensen SE, Westlake DWS, Wolfe S (1985) J. Antibiot. *38*: 263

Author Index Volumes 1–39